WHEN Oil PEAKED

WHEN **Oil** PEAKED

Kenneth S. Deffeyes

HILL AND WANG
A division of Farrar, Straus and Giroux
New York

HILL AND WANG
A division of Farrar, Straus and Giroux
18 West 18th Street, New York 10011

Library of Congress Cataloging-in-Publication Data
Deffeyes, Kenneth S.
    When oil peaked / Kenneth S. Deffeyes. — 1st ed.
        p.   cm.
    Includes bibliographical references and index.
    ISBN 978-0-8090-9471-4 (hardcover : alk. paper)
    1. Petroleum reserves—Forecasting—Popular works.
  2. Petroleum industry and trade—Forecasting—Popular works.
  3. Mines and mineral resources—Popular works.   I. Title.

TN871.D395 2010
333.8'232—dc22

                                        2010002437

Designed by Cassandra J. Pappas

www.fsgbooks.com

For Angus,
from KSD

It is difficult for people living now, who have become accustomed to the steady exponential growth in the consumption of energy from the fossil fuels, to realize how transitory the fossil fuel epoch will eventually prove to be when it is viewed over a longer span of human history.

—M. KING HUBBERT, *Scientific American*, 1971

Contents

Preface

There is an ongoing debate between "peak-oil" people, those who think that we have already found most of the world's oil, and their opponents, who insist that tweaking the economic system will enable us to find lots of additional oil. Colin Campbell, one of the peak-oil leaders, has repeatedly referred to Michael Lynch as "that flat-earth economist." For a long time, I thought that it was just an insult. (Lynch never claimed to be an economist.) Finally, I understood. On an extended Earth you can have an infinite amount of oil. On a round earth, there are limits.

Most of the peak-oil people credit the American geologist M. King Hubbert (1903–1989) for making the earliest solid analysis of the problem. His followers are sometimes called "Hubbertians." The opponents are "cornucopians," named for the traditional horn of plenty. At the moment, the intellectual debate remains unsettled. Out in the real world, however, the story is unfolding; the contest is being decided in the marketplace. I've been a Hubbertian since the late 1950s. I'm a geologist, with no professional expertise in economics or politics. My arguments come largely from the viewpoint of earth science. The purpose of this book is to update the peak-oil debate and then to extend the coverage to include other resources. Are we next going to see "peak copper" or "peak phosphate"?

In my first oil book, *Hubbert's Peak* (2001), I followed Hubbert's methods and predicted that world oil production would peak in the year 2005. Currently, the U.S. Energy Information Agency shows 2005 as the peak year, with 2008 in second place. Blow the trumpets! Hubbert's method actually works!

The Round Earth

On a spherical Earth, there is a fixed surface area that we can explore for oil. World oil production increased rapidly from the first wells, around 1859, up to the year 2005. From 2005 onward, oil production has shown no growth. This, of course, is the story that is unfolding before our eyes. Will delays and cancellations brought on by the 2010 BP blowout in the Gulf of Mexico leave 2005 as the peak year? Stay tuned, this is your future.

Obviously, we have been handed a major financial crisis. Initially, Barack Obama's staff was heavily focused on thawing out the financial system. I once had a computer with a clickable item called "last good menu." It was a way to go back to the time before the computer crashed. The White House staff seems to be looking for a reset button to return us to 2003. Of course, it won't work. Everybody won't get their jobs back, or their houses; world oil production will not increase. The invisible hand of economics is now clenched into an invisible fist, pounding the world economy down to fit the resource base.

To his credit, President Obama has stated that the energy crisis and health costs are major important items. Crude oil prices had already doubled before the Bear Stearns collapse in March 2008. The Dow Jones Industrial Average peaked even earlier, on October 5, 2007. It seems obvious to me that the oil problem is the disease and the financial recession is a symptom. Oil production ceased growing in 2005, as I had predicted four years earlier. In May 2005, oil was priced at 44 dollars per barrel, and in July 2008, crude oil reached 147 dollars per barrel. Oil production in July 2008 was 0.2 percent higher than in May 2005. That wasn't exactly what we learned in Econ 101: Tripling the price brought out only a trivial amount of additional oil. The chant of "Drill, baby, drill" at the 2008 Republican National Convention had a hollow ring to it.

Economists and geologists have a long-standing disagreement about the availability of resources. I worry that the emphasis on economic growth stems from their operating the U.S. government as an enormous Ponzi scheme. I'm not alone in that concern. I looked up "Social Security ponzi" on Google and got 310,000 hits.

It's just a geological accident that oil caused the problem. It might have been phosphate instead. OPEC would have been the Organization of Phosphate Exporting Countries, with exports from Morocco as number one and the United States as number two. A phosphate crisis would have been enormously more disruptive than too little oil.

Although I grew up in the oil fields, I barely recognize the oil industry as it is today. When I was young, the major oil companies (the Seven Sisters) dominated exploration, production, refining, and marketing. The majors generated wealth by finding and owning giant oil fields. Today, national oil companies dominate most of the oil-productive areas. Whereas the major oil companies used to operate large corporate research laboratories, innovation now comes largely from service companies like Schlumberger and Halliburton. The Seven Sisters are still around, although some of them have been absorbed through mergers. In a sense, each major now consists of

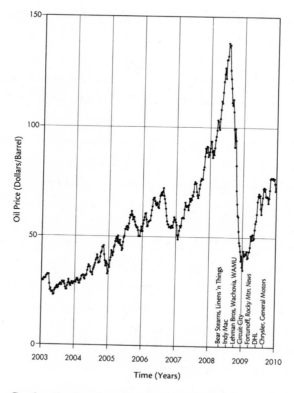

Crude-oil prices (uncorrected for inflation) and the
fall of the American economy. The Bear Stearns
collapse is usually regarded as the first major incident
at the beginning of the meltdown. The price drop in
early 2009 did not magically transport us back to the
year 2004. (© Steve Deffeyes)

skilled professionals with a merchant bank attached. "Will work for
a small share of the oil."

Although M. King Hubbert is regarded as the patron saint of
the peak-oil crowd, his methodology and his conclusions are still
the subject of criticism and skepticism. It's important not to claim
that Hubbert is infallible; I point out several flaws in his work in
this book. Of the many Hubbert critics, Richard Nehring published
one of the most detailed analyses, and his critique is backed by data.

I explain in chapter 2 how Nehring's analysis runs afoul over small-sample statistics.

An original contribution in chapter 1 of this book is an attempt to supplement the logistic bell-shaped curve (the one that Hubbert used) with a similar justification for the Gaussian bell-shaped curve. The Gaussian curve has a wider top and narrower tails. An evaluation of several different bell-shaped curves, using several data sets and several fitting methods, was done in 2007 as a Princeton senior thesis by Benjamin Steiner. Steiner's conclusion was that all combinations of math, data, and methodology gave essentially the same results for the next ten years.

Oil prices are a subject of never-ending interest. Chapter 4 contains a graph of the total annual price paid for crude oil as a percentage of the gross domestic product. Oil costs during the 1990s were less than 1 percent of GDP; those years brought us the SUV gas guzzler. Oil costs exceeded 4 percent of GDP in 1981 and again in 2008. Both of those price spikes caused extensive damage to the U.S. economy.

Oil prices have a selective impact on various countries. In particular, Iran and Venezuela need a price above seventy dollars per barrel to support their national adventures. In most of the world, low oil prices (and the possibility of even lower prices) have forced the postponement or cancellation of a substantial number of oil projects.

There are some bits of good news from the oil patch. The question is whether they are just a small amount of good news embedded in a larger mass of bad news. The bad news is that the supergiant oil fields that supply much of our oil are getting older and declining in production. For oil, I suspect that the good news is not good enough; it is not a game changer. For natural gas, the good news *might* be the opening of an era of increased supply.

The best oil news comes from offshore Brazil. Petrobras, the Brazilian national oil company, has made a string of remarkable new discoveries of light high-grade oil, possibly the biggest discoveries since 1976. However, the wells are very deep and very expensive to drill and to operate. If the price of oil is too low, the deepwater Brazilian production will be zero. In a neatly ordered economic world, the crude-oil price would settle just above the level that makes the Brazilian offshore fields viable. Unfortunately, at their best, the Brazilian fields will not offset the production declines elsewhere in the world.

For natural gas, there is a new ball game. We can state it in a pithy seven-word sentence: "Mature oil source rocks will yield gas" (this mantra is explained further in chapter 9). Suddenly, whole new provinces of natural-gas fields have become available. On an equal-energy basis, about 6,000 cubic feet of gas equal a barrel of oil. Because natural gas is a cleaner-burning fuel than oil, we would expect natural gas to command a premium price. Not so. When oil is selling for $68 per barrel, dividing the oil price by six gives an energy-equivalent natural-gas price of $11.30 per 1,000 cubic feet. If natural gas is actually selling for $3.60, that means it is available for one-third of the energy-equivalent price of oil. My interpretation is that today's low natural-gas price results from the (real or imagined) extent of potentially large new gas supplies.

On a time scale somewhere between one hundred and three hundred years, our civilization has to come around to sustainable and renewable resources. Most energy will be, directly or indirectly, solar. (Wind farms, hydroelectric dams, and biofuels are indirect forms of solar energy.) We don't have to go solar next week, but we don't want to end up in any dead ends on the way to a solar future. In particular, even the enhanced natural-gas supply might not carry us all the way over to the solar Promised Land.

Among the available energy sources already identified are uranium and coal. Both are politically unpopular. However, as we run out of alternatives we may have little choice. For uranium, acceptably safe reactors have been engineered and the radioactive waste disposal problem was already solved in 1955, as discussed in chapter 7. Also in chapter 7 is a generalization of the uranium roll-front deposits to a variety of other metals.

Coal was, of course, the fuel that originally powered the Industrial Revolution. Several new evaluations of the world coal supply have appeared recently. Two centuries of experience led to new methods for finding and extracting coal. The problem is exploiting the coal without fouling Earth's atmosphere. An updated version of the 1870s technology for coal gasification may offer an acceptable way of exploiting coal (chapter 6).

Burning any of the fossil fuels adds carbon dioxide to the atmosphere. The split between the peak-oil freaks and the climate-change folks is more profound than we like to admit in public. Warnings about climate change usually carry the implied message that any change is going to be disastrous. Geologists are aware of giant changes in climate through geologic time. Only seventeen thousand years ago, New York City, Toronto, and Stockholm were buried beneath a mile of ice. (In a four-billion-year history, seventeen thousand years is the day before yesterday.) In chapter 8, I evaluate both large-scale and small-scale climate-modification schemes.

Recently, I conducted a family energy survey around the dinner table. Miles driven, natural-gas bills, electric bills, and grocery prices. To my surprise, the largest energy cost was at the grocery store. Modern agriculture is highly dependent on fossil fuels. At the same time, converting corn, sugarcane, and oilseeds into biofuels reduces the human food supply, raises food prices, or both. Locally grown food is not just about organic fertilizer or fresher crops; it is an alternative to flying fresh fruit up from Chile during the northern winter. Of course, there will be disputes over water usage and farmland availability. It's not that water and land aren't important; I simply don't know enough about them to hold strong opinions.

Coal-powered Mississippi River steamboats were an American icon during the nineteenth century. (iStockphoto)

Transportation efficiency will be increasingly important. Gasoline and diesel fuel, because of their portability, are the leading users of crude oil. In chapter 11, I present an overall view of transport efficiency. One surprise emerges: Oceangoing ships and river barges are the most energy-efficient methods for transporting materials. River traffic takes us back to an earlier history. When Samuel Clemens needed a pen name, he chose the leadman's cry for water two fathoms (twelve feet) deep: "mark twain."

A section in chapter 13 explores the possibility of using "critical damping" (borrowed from engineering techniques) to reduce volatility in the stock market.

WHEN Oil PEAKED

ONE

Bell-Shaped Curves

ubbert's original analysis, in 1956, was about U.S. oil production. That's not a matter of patriotism; the United States is the most intensely drilled area in the world. He began with two different estimates for the total amount of recoverable oil beneath the United States. The two estimates came from two senior petroleum geologists, Wallace Pratt and Lewis Weeks. Hubbert said that a graph of annual U.S. oil production versus time would have a bell shape, with a fast growth at the beginning, a curved peak in the middle, and a tailing off at the end: in round numbers, about one hundred years from the start to the peak and one hundred years of decline from the peak until the last well runs dry. There were two important constraints:

- The area under the bell-shaped curve had to match the estimate for the total U.S. oil.
- The early part of the curve, 1859 to 1956, had to fit to the historical U.S. oil production.

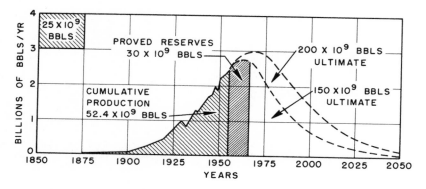

M. King Hubbert's initial 1956 graph. His original caption read: "United States crude-oil production based on assumed initial reserves of 150 and 200 billion barrels."

Hubbert then had two curves, one for each of the estimates, one from Pratt and one from Weeks. The lower estimate, from Pratt, of 150 billion barrels had its peak around 1965, and Weeks's more optimistic estimate of 200 billion barrels peaked around 1970. U.S. oil production did peak in 1970, and Hubbert moved up from being a crank to a folk hero. However, there are two substantial lessons from this episode:

- Hubbert had no way of knowing which estimate was correct. Most analysts have given him credit for a direct hit when using the higher estimate. (Long after the fact, I reworked his data; he had no grounds for rejecting the lower estimate using the production data up to 1956.)
- Critics of Hubbert's method have repeatedly said that *all* Hubbertian analyses require a prior guesstimate of the total oil. By 1960, the U.S. oil-production history was far enough along for Hubbert to dispense with the guesswork and sweat the total amount of oil out of the data. None of the modern analyses involve guessing the total recoverable oil.

Even with sufficient data, there are choices to be made. There are several different kinds of bell-shaped curves. The best-known bell curves are symmetrical. The history before the peak is a mirror image of that happens after the peak. (Later, I'll explain the arguments for accepting symmetry.) However, some cornucopians have challenged the symmetry arguments, hoping for a long and very gradual decline after the peak. I have a warning: Around 1980, I worked alongside the Princeton statistics department analyzing oil problems. The statisticians found an equation for a bell-shaped curve that was not symmetrical; the later downside could be either more gradual or more abrupt than the upside. They dumped the equation and the oil-production numbers into the computer and out came a disaster. The statistically best-fitting result was an abrupt crash down to zero production soon after the peak.

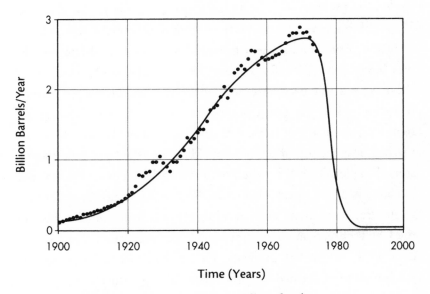

In a 1979 study, a bell-shaped curve was allowed to be nonsymmetrical; the downside was not required to be the mirror image of the upside. The computer's best fit to the U.S. crude-oil production showed a gigantic crash around 1980. Obviously, it did not happen. (© Steve Deffeyes)

The Gaussian Versus the Logistic

By far the most famous bell-shaped curve is the Gaussian. It is used so frequently that it is also known as the "normal" curve. Unfortunately, that implies that all other bell-shaped curves are "abnormal." Over my workshop desk is a framed (pre-euro) German ten-mark bill, with a portrait of Carl Friedrich Gauss (1777–1855) and a graph showing the Gaussian bell-shaped curve. Gauss made important discoveries all across the field of mathematics; his Gaussian curve for statistics was a small part of his total contribution.

In his original 1956 paper, M. King Hubbert used a different bell-shaped curve called a "logistic" curve. Typical of Hubbert, he did not explain his reasons until 1982, when he was seventy-eight years old. Hubbert was a difficult character. Here's an example: When Hubbert and I were both working at the Shell research lab in Houston, a graduate student from Cal Tech gave a talk because he was being considered for a job at the lab. The student's doctoral thesis was a microscopic examination of a sample of rock about 6 inches square. He put a photograph on the screen showing the rock face, about 6 feet high, with the location of the sample square marked near the top of the outcrop. Hubbert asked a question about something lower on the photograph and the student said that it was outside his thesis area. Hubbert was furious. The student did not get a job offer from Shell. Further, Hubbert contacted the chairman of the Cal Tech geology department with the suggestion that the student not be allowed to graduate. The student eventually did graduate, but all of us learned a lesson about dealing with Hubbert.

The logistic curve that Hubbert chose is, at first glance, similar to the Gaussian. The rounded top of the Gaussian is wider than the logistic; in compensation, the Gaussian has narrower tails on the far

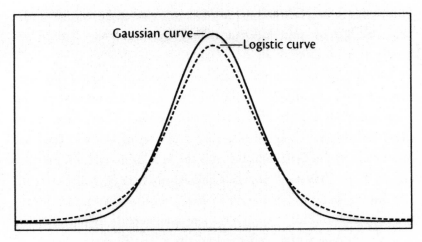

Comparison of the Gaussian and the logistic curves, with the same area beneath both of the curves. (© Steve Deffeyes)

left and right. For me, the choice between the two came up early. Around 1962, I considered leaving the Shell research lab and taking a university faculty appointment. Should I bet my career on changing my way to earn a living? A major consideration was Hubbert's 1956 prediction about U.S. oil production. If Hubbert was right, by the year 2000, when I was scheduled to retire, the U.S. oil industry would shrink to about half of its peak size. I didn't want to live in Algeria or Nigeria; I'm an American.

As is typical of scientists, I keep two mental lists. One list is scientific conclusions that I have read about and seemed reasonable; they are filed under "probably okay." My other list is shorter; it is problems that I have worked with on my own terms. Before deciding to switch from a research lab to teaching, I needed to work directly from the raw data. At the Shell lab, all I had to do was walk down one floor and ask Hubbert's assistant for a photocopy of the raw U.S. oil-production data. Hubbert had used the logistic curve; I decided to try the Gaussian. I got essentially the same results as

Hubbert: a U.S. peak around 1970 and a reduction to half of the peak size when I retired. A year later, I was a professor at the University of Minnesota.

In 2000, happily retired, I returned once again to the logistic-versus-Gaussian outlook on oil production. By a narrow margin, the Gaussian made a better fit to the historical data. However, most of the difference between the Gaussian and the logistic was the narrower tails of the Gaussian. Because the U.S. oil industry grew very rapidly from 1859 to 1910, the Gaussian made a slightly better fit. But are we going to bet the ranch based on the happenings during the early days of oil? There is a deeper issue here. Is there some reason why Hubbert preferred the logistic curve? Does the Gaussian have a hidden flaw? Or, as some critics believe, are both of the curves somewhat off?

The Logistic Versus the Gaussian

The logistic curve has a logical story attached to it. At least for oil production, the Gaussian has lacked a logical justification. At the end of this section, I'll remedy that problem by introducing my homemade story.

We can begin by plotting the U.S. oil production on two similar, but not identical, graphs. The upper graph is contrived to plot a straight line if the production history follows a logistic curve. The lower graph gives a straight line for a Gaussian history.

On these graphs, P is the annual crude-oil production and Q is the cumulative production, beginning with the Drake well of 1859. (In this racket, Q stands for *cum*ulative.) P' is the increase (or decrease) in production from one year to the next. (In nerdish, P' is the time derivative dP/dt.) The points a and b on both graphs in-

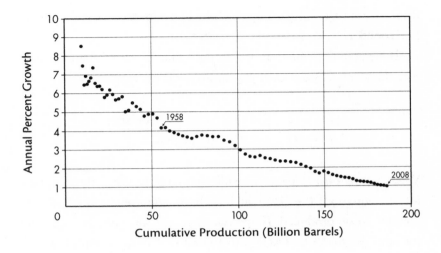

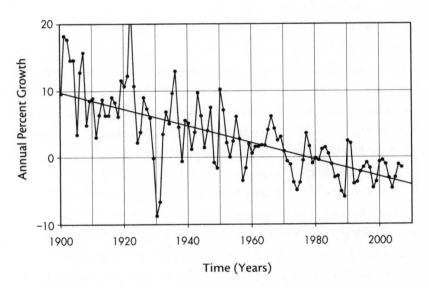

Comparison of the linearized graphs for the logistic (*top*) and the
Gaussian (*bottom*) for U.S. oil production. The Gaussian graph looks
more scattered because year-to-year variations become emphasized,
whereas short-term variations are smoothed in the cumulative logistic
graph. (© Steve Deffeyes)

dicate where the best-fitting straight line hits the vertical and the horizontal axes.

If we take the equation of a straight line ($Y = a + bx$) and substitute the symbols from each graph, we get these equations:

$$P = a \left(1 - Q/b\right) Q \quad \text{(Hubbert, logistic)}$$
$$P' = a \left(1 - t/b\right) P \quad \text{(Gaussian)}$$

What jumps off the page is inside the parentheses in the Hubbert logistic equation: $1 - Q/b$, which is the fraction of the total oil that remains to be produced. If, instead of producing, you are exploring for oil, $1 - Q/b$ is the fraction of the oil that remains undiscovered. Early on, when most of the oil hasn't been found, exploration is really easy. Toward the end, when most of the oil has been discovered, exploration is not very rewarding. A fellow who sees a friend fishing at the edge of a pond shouts, "How's the fishing?" The reply comes back: "The fishing is great. The catching stinks." That's the present-day story about oil exploration, the catching stinks. A few years ago, Chevron ran a series of ads saying that we are burning two barrels of oil for every new barrel we find. Today, we may be burning five barrels for each newly discovered barrel.

The logistic equation was developed in 1838 by Pierre-François Verhulst to describe population growth. Imagine a freshly bulldozed building lot that is left untouched for a while. A few weeds grow first. Eventually plants cover the entire lot and the population stops increasing. The logistic equation says the rate of plant growth depends on the fraction of the space that is still open for colonization. It was this behavior that Hubbert utilized to say that the ease of finding oil depends on the fraction of the total oil that remains undiscovered.

In 1967, the meaning of the logistic equation was refined by Robert MacArthur and E. O. Wilson. They pointed out that the two constants in the equation, a and b, were related to two different strategies. Weeds (and mosquitoes during the brief Arctic summer) are dominated by a, although they used the letter k in their equation. Natural selection for large a leads to reproductive strategies involving rapid maturity, a large number of offspring, and dispersal strategies for scattering the offspring. (After a hike, you find weed seeds stuck in your socks.) At the other extreme, selection for b, which MacArthur and Wilson called R, happens in mature and stable environments. Elephants are the extreme example: large body size, few offspring, the rarity of twins, long life, long gestation period (22 months), and a major investment in parental care. In the case of elephants, "parental care" involves the mother and a cluster of sisters and aunts.

In the oil business, there is a rough analogy. When a new area starts being explored, there are boomtowns and drilling rigs all over. In a mature area, there may be a few ultradeep wells being drilled.

Hubbert's logistic model has a nonbiological peculiarity: No barrel of oil ever dies. Although we may burn up the oil while driving the kids to Yellowstone, the oil is listed permanently in the cumulative roster of produced oil. To Verhulst, as well as to MacArthur and Wilson, population growth was an excess of births over deaths.

On the logistic curve, the initial growth stage is the mirror image of the later decay period. There are two ways to reach that conclusion; you are welcome to use either or both:

- If the logistic graph shows a straight line, then the bell-shaped logistic curve will be mirror-image symmetrical.
- If oil-finding success depends exclusively on the fraction of currently undiscovered oil, then the bell-shaped logistic curve will be mirror-image symmetrical.

It is important to notice that either of these conditions lock in mirror symmetry. Some of Hubbert's critics have claimed that mirror symmetry has to be proved by additional observations.

When we turn to the Gaussian, the first condition will work. If the Gaussian linear production graph shows a straight line, the Gaussian bell-shaped curve will be symmetrical. However, for the second logistic argument (discovery depends on the undiscovered fraction) there is no equivalent Gaussian story. Why is the story important? Philosophers may list it somewhere under "epistemology," but any used-car dealer knows it as a "sales pitch." It makes us feel better if there is an equivalent of Kipling's *Just So Stories*.

Here's an interesting story, but one that doesn't apply to petroleum: The vertical axis of the Gaussian graph is P'/P. Since P' is the annual change in production, then P'/P is an interest rate. If you have a savings account yielding 4 percent, the balance in your bank account gets multiplied by 1.04 each year. Now let's look at the situation where the interest rate goes down linearly with time. At first the interest rate is high and your account grows rapidly. Gradually and steadily, the interest rate goes down and your account grows more slowly. Then when the interest rate goes to zero, your bank balance ceases to grow. I know, I know, by then you would close your account and move the money elsewhere, but we're doing a thought experiment. What if you left the account alone as the interest rate moved into negative territory? Your account would shrink slowly at first, and then more rapidly. Eventually, the balance in your account would become very small. Here's the surprise: The balance in your account over time would draw a perfect bell-shaped Gaussian curve.

Can we use the savings account analogy to describe the oil business? If God were upstairs, keeping the books and doling out oil discoveries at a linearly decreasing annual rate, then we would have a story. It doesn't seem to me that God would want to act like an enlarged version of Bank of America. We'll have to look elsewhere.

Quite separate from the bank account analogy, there is another way to generate a Gaussian curve. It's called the central limit theorem, which sounds impressive; you could try to work it into your party conversation. When some guy is talking loudly about housing prices, ask him whether the central limit theorem explains it. The central limit theorem is easiest to understand through an analogy. Let's say that we have a rifleman shooting at a target. The bullet can hit the target to the left or right of the center for a number of reasons:

- There is a gusty wind blowing from left to right.
- The rifleman can't hold the rifle absolutely steady.
- Some of the bullets are a bit heavier, or lighter, than the others.
- The sights on the rifle are a bit out of adjustment.
- Some of the bullets are backed by a little more, or less, gunpowder.

Each of these variations needs to be independent of the others. There is no Murphy's Law saying that the rifleman always selects a lighter bullet on windy days. The central limit theorem says that adding up a bunch of these independent variations gives rise to a bell-shaped Gaussian distribution of bullet holes to the right and left of the center.

I'll put on my Rudyard Kipling hat and try to develop a Gaussian story for oil exploration. Be warned: Kipling and I made up these stories as we went along. There are obvious clues and subtle clues for oil discovery. If there are a number of clues, and if they are independent of one another, the central limit theorem would then get us a Gaussian history of oil discovery. We need at least five or six independent attributes, but the more the merrier. There are eight in the list below. The idea being tested says that the typical, average, plain-vanilla oil field would have been discovered in 1963. However, each of the eight independent conditions could advance or delay an oil field's discovery before or after 1963. These are not on-off toggle switches; there is every gradation from obvious properties that lead to early discovery, to subtle hints that don't drastically change the discovery date, to situations that substantially delay finding the oil field. In the extreme, the time leads or time lags can be as much as one hundred years.

- **Surface Indications of Oil or Gas** Early discovery: oil seeps, tar pits, or natural-gas leaks at the surface, as in the Bible (the bush that was burning and not consumed, a pillar of cloud by day and of fire by night, and the fiery furnace in the Book of Daniel). Late discovery: oil fields beneath nonleaky subsurface salt layers, as in offshore Brazil.
- **Geologic Structure** Early discovery: visible evidence of structural domes in the surface rocks, with names like Circle Ridge or Racetrack Valley. Late discovery: oil traps such as deeply buried fossil reefs, with no surface expression at all.
- **Geophysical Exploration** Early discovery: gravity surveys over salt domes in the U.S. Gulf coast. Late discovery: Difficult interpretation of reflected seismic waves sent through the salt layer in the deepwater offshore U.S. Gulf coast and offshore Brazil.
- **Location** Early discovery: Oklahoma. Late discovery: Bhutan, where we haven't even started.
- **Downhole Well Logging** Early discovery: high electrical resistance due to oil or gas instead of salt water, developed by the brothers Schlumberger in the 1920s. Late discovery: oil reservoirs with marginal electrical resistance, requiring skillful interpretation.
- **Politics** Early discovery: United States and czarist Russia. Late discovery: central South China Sea (ownership disputes).
- **Clustering** Early discoveries: "fairways" and "trends," as in the onshore Texas Gulf Coast. Late discovery: isolated small oil fields, such as in Nevada.
- **Technology** Early discovery: cable-tool drilling copied from

water-well drillers. Late discovery: horizontal wells coupled with multistage hydrofracturing.

For each oil field in the world, we start with the year 1963 and subtract the advances or add the delays for each of the attributes to get the likely discovery year. The central limit theorem assures us that we will get a Gaussian distribution when we add up the discoveries for each year.

It isn't perfect. The Gaussian curve extends from the infinitely old to the indefinite future. My partial answer is that some of my Choctaw ancestors gathered oil from oil seeps. "Asphalt" and "naphtha" are ancient words from the eastern Mediterranean. The oil business did not begin with the Drake well.

Basically, I think it works. With a little tuning up, adding independent advances and delays would place the Gaussian alongside the logistic as plausible stories for the oil history. I know, it makes you nervous that I just now made up the story. It makes me nervous, too. But these ideas have to start somewhere. You were here when the Gaussian got its curve.

TWO

Nehring's Critique

S tarting in 1956, lots of critics tried to talk their way out of the bad news implicit in Hubbert's original paper. Although his prediction of the U.S. oil-production peak turned out to be essentially correct, criticism of Hubbert's methods has continued. A common complaint is rather shallow: "Most of those predictions turned out to be wrong." In my opinion, the best of the critical analyses was published in 2006 by Richard Nehring. In three articles in the *Oil & Gas Journal*, Nehring made a test using detailed data for the San Joaquin Basin in California and the Permian Basin of West Texas and New Mexico. His conclusion is that Hubbert et al. have been blindly following a flawed model. My reaction is that the Hubbertian approach has evolved over the years. There were misconceptions in the early method, some of them introduced by Hubbert himself. The purpose of this chapter is to review the current doctrine of Hubbertarianism and then to examine Nehring's numerical tests.

Nehring's original claim to fame was a thorough catalog of the world's giant oil fields, originally published in 1980. At a time when we had a very limited overall view of the world oil situation, Nehring's compilation was enormously valuable. In the first of Nehring's 2006 papers, he asks, "Can we predict when world oil production will peak?" Of course, I claimed that my answer was all over the newspapers and showing on the filling station pumps. In 2005, world oil production stopped growing, and a bidding war ensued for the remaining oil. We watched the price shoot from $30 to $147 per barrel, and financial chaos descended upon the world economy. The huge price increases encouraged oil producers to market every possible barrel, but world crude-oil production never went above the 2005 level.

Hubbert clarified his methodology in a 150-page paper, published in 1982. Hubbert finally gave his reasons for preferring the logistic curve over the more familiar Gaussian; he included a derivation of the mathematics using differential equations and he gave a single example of what is now sometimes called the "linear Hubbert" graph (P/Q versus P). In his *O&GJ* papers, Nehring does not explicitly use the 1982 Hubbert publication. In contrast, I dug out goodies from Hubbert's 1982 paper when I was writing *Hubbert's Peak* and *Beyond Oil*.

No surprise—Hubbert did not turn out to be infallible. There were two major problems plus one small but glaring boo-boo. The first big problem is a dual meaning of "discovery," and the second problem concerns the identical shape of the discovery and production curves.

The Discovery of "Discovery"

In all of his papers, Hubbert carefully defined cumulative "discovery" for a given year as the total amount of oil produced up to that year plus the known underground reserves as of that same year. Unfortunately "discovery" was already embedded in ordinary English and in oil country slang as the date of the first well in a field: the discovery well. My reading of the *O&GJ* papers gives me the impression that Nehring uses the original Hubbert definition, but the other "discovery well" meaning keeps trying to emerge. I used the same definition in *Hubbert's Peak* and I mistakenly followed Hubbert's lead and reported that the "discovery" curve was identical in shape to the production curve but was simply displaced eleven years earlier.

Using known reserves was Hubbert's way of utilizing what economists call a leading indicator. Unfortunately, after Hubbert's time, the reporting of reserves became increasingly unreliable. Colin Campbell pointed out in 1997 that most, but not all, OPEC members abruptly doubled their reserves during the middle 1980s. Campbell wrote that it was a response to the change in OPEC allowable production rates from a dependence on production capacity to a dependence on reserves. In the United States, overly optimistic reserve estimates led the Securities and Exchange Commission to introduce a rather tight definition for reported reserves. In 2004, Royal Dutch Shell was forced into a major reorganization because of overreported reserves. *Oil & Gas Journal* publishes a widely used statistical summary at the end of each year, but *O&GJ* clearly states that a questionnaire is sent each year to the oil companies, including the national oil companies, and the year-end summary comes from the questionnaire replies. If a national oil company does not reply, *O&GJ* simply repeats the previous year's estimate.

An additional hazard involving reserve reports comes from newly discovered oil fields. Three-dimensional reflection-seismic surveys, especially at sea, are an important improvement over the earlier two-dimensional methods. Practitioners of the art sometimes think that they can see the outline of an entire oil field after the first successful well is drilled. It is not uncommon to read news releases that estimate ten billion barrels of oil on the basis of a single well. Years later, the field sometimes turns out to be much smaller or not economic at all.

For me, the bottom line is that national reserve estimates contain very little useful information. Within the United States, the regulated reserve estimates are probably too low, but at least the reserve figures give the relative sizes of the different companies. For several years, Matthew Simmons, founder of a major investment bank, has been calling for international field-by-field verifiable reserve estimates. I'd love to see that, but I'm not holding my breath. Until then, I'm not about to bet the ranch on any analysis that depends on the reported reserves.

Of Time and the Hits

The confusion between Hubbert "discoveries" and ordinary field discoveries can be solved by retaining Hubbert's definition of discoveries (cumulative production plus reserves) and opening a new category called "hits." The hits category credits all the oil that will likely be produced from an oil field to the year of the first well in the field. Having three categories—hits, "discoveries," and production—solves the intellectual problem at the expense of creating a grammatical problem. When I am talking about it, I have to stop each time I start to say "discovery well." Maybe instead I should have renamed the Hubbert "discoveries" as "points on the scoreboard." For now, I'll leave Hubbert's definition unchanged as a quaint anachronism, but I will keep "discovery" in quotation marks to flag it as nonstandard.

A major conceptual mistake in Hubbert's work was presuming that his "discovery" curve had the same shape as the production curve but was simply displaced in time. Nehring illustrates the relationship in Figure 1 of the first of his 2006 papers in *O&GJ*. My preferred alternative says that the three curves—hits, "discoveries," and production—all have to start on the same year. The areas under the three curves have to be identical; we are just sorting the same barrels in three different ways. The resolution involves making the hits curve narrower and higher, Hubbert's "discovery" curve is intermediate, and production is wider and lower, as shown in the following graph.

The hits have a special significance. After the first successful well in a field is found, you don't get to find it again. (See: I wanted to write *discover* instead of *find*.) In a few instances, several smaller oil fields eventually grew together into one major field. Around 1980, the U.S. Department of Energy had a one-person project to remove

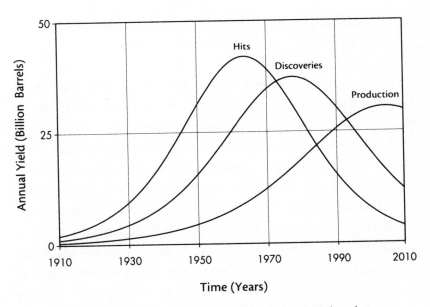

World crude-oil logistic curves for "hits" (crediting the oil to the year of the first well in the oil field), for "discoveries" in the Hubbert sense of crediting the oil to the year it was added to the list of known reserves in the ground, and "production" in the ordinary sense of oil brought to the surface and marketed. (© Steve Deffeyes)

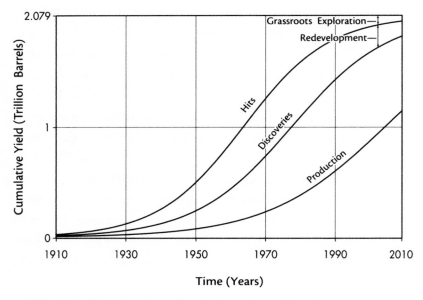

The same data as on the graph on page 21, but shown as cumulative curves added up from the beginning. (© Steve Deffeyes)

those duplicates from the first-well statistics. I don't know what became of the study (if any reader knows of a copy of the results, please send an e-mail to deffeyes@princeton.edu).

In addition to the bell-shaped curves for the three categories, we can plot the same data as cumulative curves. We just add up each curve from the beginning and eventually (on the right) the three cumulative curves have to top out at the same level. There are some interesting consequences. The hits curve is 95 percent of the way to the ceiling. That means that the oil fields we have already found contain 95 percent of all the oil we are ever going to find. This is an enormously unpopular result. For starters, it means that the major international oil companies are not going to generate additional wealth by finding new giant oil fields. For the larger world economy, it means that "business as usual" will not prevail. For us little people, it means that bidding wars (we hope not shooting wars) for the remaining oil will continue.

The middle curve, Hubbertian "discoveries," comes with an attached warning: The reserves estimates are not highly reliable. Taken at face value, the gap between "discoveries" and hits is about three times larger than the space between hits and the ceiling. The message is that finding additional oil in our existing oil fields is likely to be much more rewarding than grassroots exploration for new fields. So how do these rewards come about?

During the 1960s, Shell Oil had a tiger team of specialists in California who would be pulled from their regular assignments whenever an oil field or an oil company was up for sale. The team typically was made up of a petroleum geologist, a well-log analyst, a petroleum engineer, and others as needed. Their task was to ferret out productive layers in the oil field that had been overlooked during its original development. The original 1920 invention that founded the Schlumberger dynasty was the downhole measurement of electrical conduction. Rocks saturated with salt water are good electrical conductors; oil- and gas-saturated rocks are poor electrical conductors. The Schlumberger company called itself "the eyes of the oil industry." In some instances, the eyesight was a little blurry. There were zones of intermediate electrical conductivity; typically these were ignored before 1947 and, sometimes, even later.

The 1947 breakthrough was a quantitative calculation pioneered by Gus Archie, who was working for Shell Oil Company. Knowledge of the rock porosity, the electrical conductivity of the water, and the measured downhole electrical conductivity allowed the well-log analyst to compute the percentage of the rock pore space that was filled with oil or gas. The break point is around 50 percent oil and 50 percent salt water. More than 50 percent oil, the rock would produce oil and some water. Below 50 percent oil, the rock would flow almost all water. Here is where the skill comes in: A small difference above or below 50 percent is worth a huge amount of money.

Several of us who worked in Gus Archie's group at the Shell research lab went our different ways in the mid-1960s. While I chose the academic route, Bob Sneider made an unusual choice. He assembled his own tiger team and went into the business of acquiring existing oil fields with overlooked productive zones. He bought 46 oil fields and added 625 million barrels of oil in West Texas and the Gulf Coast. Sneider published his methods; my interpretation is that he succeeded not because he was a supertiger but because he never did anything wrong.

Bob Sneider retired from active participation in oil field acquisition, and he died in 2005. However, the large gap between the Hubbert "discoveries" and hits suggests that Texas is not unique. Schlumberger began offering "redevelopment" as one of their services: rent-a-tiger-team. In Iraq, for instance, there may be major opportunities to locate unproduced horizons within the existing oil fields. Programs are starting to appear aimed at redeveloping those fields.

World Hubbert Graphs

The graph on page 25 is similar to the linear graph on page 9 but it is for world oil production instead of U.S. production. The vertical axis is the annual production for each year, divided by the cumulative production up to that year (P/Q). The horizontal axis is just the cumulative production. After 1980, the dots form a pretty good straight line. If you are trying to predict the future, a straight line is your mother's comfort food. The most reasonable guess is that next year's dot will be on an extension of the straight line. A few years ago, I received a funny e-mail from an oilman in Aberdeen who was

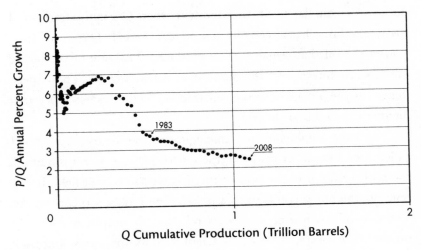

World crude-oil production, shown on a graph that turns a logistic bell-shaped curve into a straight line. (© Steve Deffeyes)

winning cash from his colleagues by betting that production would decline from the British side of the North Sea.

Notice that I have used only the production numbers. There are no reserve estimates or opinionated guesses (about eventual production) that go into making the graph. Most oil outside the United States is delivered by tanker ships, and as tankers are unloaded they are monitored for both the size of the cargo and its origin. Hint: You get the information by bribing the tax collector at the unloading dock.

The equation for the straight line on the linear graph is

$$P = a\,(1-Q/b)\,Q$$

That equation is a compact one-line statement of the Hubbert theory. In the equation, b is the place where the straight line hits the horizontal axis. (In older treatments, Q_t had the identical meaning as b.) If you wish, you can think of b as the total amount of oil produced when the last well runs dry. If so, then Q/b is the fraction of the total oil that has already been produced, and $(1-Q/b)$ is the fraction of the oil that remains to be produced. The ease of producing oil depends linearly on the amount of remaining oil. The dependence on the yet-to-be-produced fraction is the just-so story that justifies the Hubbert theory.

The same graph, and the same equation, can be used for what I called hits. Instead of crediting the oil to the year it was produced, the oil is credited to the year of the first well in the oil field. In my opinion, the hits graph is the reason why the Hubbert theory works. The ease of discovering additional oil depends simply, and linearly, on the undiscovered fraction. A scientific tradition, extending back to William of Occam around 1330, says to try the simplest explanation first. If the simplest explanation fits the observations, be happy. If it doesn't fit, you are authorized to look at something a little more

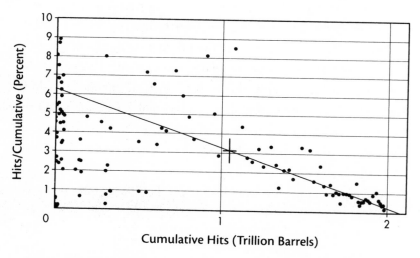

World "hits" (attributing the oil to the year of the first well in each oil field). The gloomy conclusion is that the fields that we have already discovered contain 95 percent of all the oil that we will ever find. (© Steve Deffeyes)

complex. In this case, the fit is quite good. We should be intellectually pleased, but it's bad financial news.

Wouldn't the introduction of new technology cause the straight line on the graph to take off in a new direction? In principle, it might. However, new technologies have steadily been introduced over the long history of the oil business. Technical innovations have been more than welcome, but none of them have been game changers that abruptly introduced a new era. Predictions from Cambridge Energy Research Associates, from the U.S. Geological Survey, and from ExxonMobil all require a sudden upward dogleg in the graph. My feeling is that the world oil industry is a large system and there is no current news big enough to point the system in the direction of increased production.

If the Hubbert theory is a statistical statement about finding oil, then the theory should be useful only for areas large enough to generate smooth statistics. Nehring's two examples in his 2006 papers are the Permian Basin of West Texas/New Mexico and the San Joaquin Basin in California. Although there are a number of oil fields in each of Nehring's areas, think for a moment about an area small enough to contain only one oil field. The Hubbert analysis wouldn't work because the single oil field would be developed— drilled up—in about ten years, and the ensuing production might tail off over the next hundred years. It wouldn't give a bell-shaped, symmetrical curve. In the case of the Permian Basin, the early discovery of the Yates field was a huge initial boost. (As of 1929, there were ten wells at Yates with one-hour flow tests larger than one hundred thousand barrels per day.) Also, for the Permian Basin, Nehring doesn't point out that until 1972, production rationing artificially held down production in the Texas portion, which is most of the area. Nehring's conclusion is that Hubbert's method tends to underestimate the eventual oil production. That's something we would all be happy to accept. However, it may simply be a result of the small sample size in Nehring's two areas. I cannot make an independent analysis of Nehring's data because he seems to be working from one of the private data sets that he markets.

I do agree with Nehring's interest in hindcasting. When could Hubbert have reliably made his earliest prediction? I wrote a computer program called HUBBERT.BAS that ran a Hubbert-style analysis for U.S. oil production as of each year from 1930 onward. For the early years, the annual Hubbertian estimates jumped around wildly. Around 1965, they settled down to essentially modern values. My conclusion was that Hubbert's 1956 paper benefited from some educated (and/or lucky) guesswork.

The acknowledgments at the end of Nehring's three 2006 papers contain a statement that he is no longer circulating reprints of his 1970s and 1980s papers because they underestimated future production and recovery.

And the small, but glaring, Hubbert boo-boo? In his 1956 paper, he had to rely on two expert estimates of the total recoverable U.S. oil. In his 1960 publication, Hubbert was able to dispense with the expert opinions; he could sweat the information out of the ongoing production data. However, as he progressed, two of his "revised" evaluations agreed, down to the last barrel. If each was a new and independent estimate, he should have fudged the later one a little so that it did not agree to the penny with his earlier publication.

THREE

Drill, Ye Tarriers, Drill

P redictions of future oil production are limited by the quality of the data. For some national oil companies, the oil reserves in the ground are highly protected state secrets. M. King Hubbert's original source was the annual end-of-year summaries published by *Oil & Gas Journal*. A different source is available in *World Oil*. As interest got more intense, the monthly production numbers from the U.S. Energy Information Agency came into wide use. An alternative data set comes from the International Energy Agency. All the data sources have to be used with caution. As an example, in 2004, the Canadians talked *Oil & Gas Journal* into listing the entire inventory of the Athabaska tar sands as "reserves." Traditionally, reserves were defined as the future production expected from existing wells using existing technology. Only a tiny fraction of the tar-sand area has existing mines to extract the oil.

In addition to the publicly available sources, there are private data sets. Some of these are available at prices higher than I can pay. The leading group, IHS Inc., was originally named Information Handling Services (not to be confused with IHS as a Christian religious symbol). IHS gradually acquired most of the other petroleum data sources: Petroconsultants, Petroleum Information, Cambridge Energy Research Associates, Jane's, and John S. Herold. Obviously, the major oil companies have the need and the resources to compile their own data. In general, the company data sets are not available at any price. Up to the year 2002, the ExxonMobil annual data on discoveries (what I called "hits" in the previous chapter) were released publicly, and their numbers are still the standard for years before 2002.

To add to the confusion, some information and some data leak out either on purpose or by accident. Here's a sensitive example: During a long conversation, an acquaintance told me a startling piece of information. I was not asked to keep it confidential, but it was obvious to me that it would have caused a multibillion-dollar problem if the information and its source emerged. I have not revealed the juicy tidbit.

On another occasion, I told a different acquaintance that I needed a history of Malaysian oil production because I was going to give a lecture in Kuala Lumpur. He loaded the information onto one of my flash drives. When I got home, I found that my flash drive contained a detailed history for every oil-producing country in the world. The data set was my first encounter with a history that included exploratory wells. (An exploratory well is a well more than two miles from an existing oil field; also called a "wildcat" well. He's out there a-drillin' among them wildcats.) I immediately went to the world total columns, because I knew that only a few new oil fields had been found during the last several years. I made a graph of the cumulative amount of oil found as a function of the cumulative number of wildcat wells. To my great surprise, the annual dots on the graph made an almost straight line. There was no recent drop in discoveries per well; from start to finish the success rate was close to four million barrels per exploratory well.

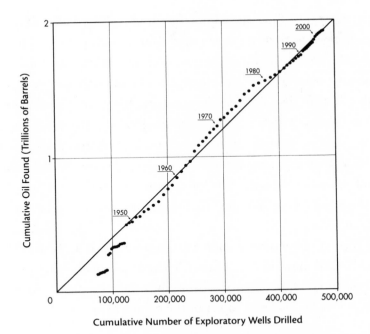

The cumulative amount of oil found depends on the number of exploratory wells (wildcats) drilled. Although the oil found is almost constant at four million barrels per exploratory well, the number of wildcats drilled per year diminished after the late 1960s. The apparent message: We are running out of good places to explore. (© Steve Deffeyes)

Of course, it wasn't four million barrels for each single well. It was more like twenty dry holes followed by an eighty-million-barrel hit. It's actually bigger than that. As early as 1956, G. Moses Knebel found that Exxon's profitability was dominated by the super-giant oil fields. (The company wasn't then called Exxon, it was Humble Oil; the company's original identity was Standard Oil of New Jersey.) Hundreds of exploratory wells go by in between discoveries of supergiant oil fields.

So why is there no bend in the oil-versus-wildcat graph? Look at the annual dots. In the heyday of exploration, around 1956, there were about nine thousand exploratory wells each year. By 2005, that had dropped to twenty-five hundred wildcat wells per year. The success rate stayed the same, we just weren't drilling very many exploratory wells. Should our message to the industry be "Drill, baby, drill"? It isn't that simple. The message that I hear repeatedly all around the oil patch is "There are no good drilling prospects out there." The oil companies, large and small, won't drill dry holes just to show their good intentions. The straight line on the oil-wildcat graph says that the standards for choosing an exploratory well have remained constant. We are simply running out of attractive drill targets.

This interpretation comes with a warning. It is based on an unpublished, unverifiable data set. There is no identification on the spreadsheets. The latest news is that the original source is no longer able to forward updated copies of the confidential data.

Inorganic Oil, Refillable Reservoirs

The standard outlook says that crude oil is generated when sediments rich in organic matter are heated at depths of a few thousand feet. The oil, being lighter than water, floats upward until it encounters a trap or reaches the surface. The trap, if discovered, is an oil field, and emptying it is a one-time process. A few people hold contrarian views: Natural gas and oil are inorganic products generated deep below the surface and upward migration will refill existing oil fields. These bits of wishful thinking are often linked together, but they are quite separate issues. I'll take the inorganic idea first and then return to refillable oil fields.

INORGANIC OIL AND GAS

Although there have long been ideas about inorganic oil and gas, the big boost for the idea arrived in 1979 when Thomas Gold began writing about it. Gold (1920–2004) was a fascinating guy. He was born in Austria and escaped to England after the German takeover of Austria. In England, he was interned as a German. After some confusion, it was realized that Jewish refugees were highly unlikely to be Nazi sympathizers and they were released. During World War II, Gold worked on radar for the British. He spent the last half of his career as a professor of astrophysics at Cornell.

Gold's strength was as an idea man, a scientist who brings up interesting new interpretations. Gold's weakness was his inability to distinguish his good ideas from his bad ideas. Gold's bad ideas were very bad indeed, but about half of his ideas were really valuable. I never wanted to ignore Tommy Gold.

Gold's original idea about inorganic natural gas was rooted in an analogy to the giant planets Jupiter and Saturn. Methane, the major component in natural gas, is also a component of the giant planet atmospheres. Presumably, the early Earth had a similar composition, although with a hard rock crust. Gradually, on Earth, photosynthesis modified the atmosphere to contain oxygen, but most of the rocks have not yet gotten the message. (Oxygen became a significant component of our atmosphere 630 million years ago, and possibly considerably earlier.) There are some sedimentary rocks, called red beds, colored by oxidized iron, but they are a small part of the overall geologic inventory. In that context, with most of the rocks retaining their original unoxidized condition, Gold suggested that methane was still leaking out of Earth's interior. An extension of the idea claimed that methane molecules could be assembled to form crude oil.

At a meeting in 1981, Gold and I gave talks about the future supply of natural gas. Rather than have a mano a mano winner-take-all debate, the meeting organizers had me talk first, saying that there is at least this much available gas. Gold then said, "Wait, there's more." Later, Gold selected a meteorite scar called the Siljan Ring in Sweden to test his ideas. Two wells were drilled deeper than 21,000 feet. (Cost: about $20 million per well.) Both wells reported tiny amounts of methane, but neither well came even close to producing commercially valuable amounts of natural gas.

The idea of inorganic oil and gas still has some enthusiastic proponents. About once a month, I get an e-mail from someone who has newly come across the idea. I suggest that they read about the gas blows from the Kidd Creek copper mine. The Kidd Creek mine, at considerable depth and in old rocks, occasionally encounters gas-filled pockets. Analysis of the gas shows abundant carbon dioxide and very little methane. It's what I would expect from degassing of the present-day Earth.

REFILLABLE RESERVOIRS

You could begin with a trip to the La Brea tar pits in urban Los Angeles. The tar pits are still active, new tar is slowly reaching the surface. As a side benefit, there is a remarkable museum with fossil bones recovered from the tar. (You are allowed to snicker a little over the name; *brea* is the Spanish word for "tar.") The La Brea tar pits are not unique; there are smaller tar, oil, and natural-gas seepages around the northern margins of the Los Angeles Basin. And world-wide, tar pits and oil seeps led to the discovery of several major oil fields.

In 1859, a group of investors considered the possibility of drilling holes next to some of the known oil seeps. They went to Benjamin Silliman Jr., who like his father was a professor of chemistry at Yale University. Silliman had analyzed oils and tar from a

The La Brea tar pits are actively bringing additional viscous natural petroleum to the surface. Ancient animals that became trapped in the tar are featured in a museum at the site. (Courtesy of the Los Angeles Public Library)

number of regions and he recognized that an oil seep near Titusville, Pennsylvania, was particularly rich in kerosene. At the time, kerosene for lanterns was the most valuable component of oil. (Gasoline for automobiles was fifty years in the future.) Silliman and the investors decided to drill a well next to the Titusville oil seep.

For their front man, the investors selected Edwin Drake, a distinguished-looking New Haven streetcar conductor. To establish his credentials, the Yale crowd mailed several letters to the Titusville hotel addressed to "Colonel Edwin Drake." I can imagine Drake bellying up to the hotel counter, "Sonny, got any mail for Colonel Drake?" "Why yes sir, Colonel, here are your letters." Drake hired a water-well driller to drill a test well, and on August 27, 1859, they made history. An oil rush ensued, and for the next fifty years the state of Pennsylvania was the world's largest oil producer. The Silliman name is remembered by Silliman College at Yale and by the important mineral sillimanite.

The conventional view (and, in my opinion, the correct view) says that organic-rich sedimentary rocks give rise to oil because temperatures at deeper than 7,500 feet are high enough to crack the larger biological materials into smaller oil molecules. (Cracking was discovered in the laboratory by—guess who?—Benjamin Silliman Jr.) The range from 7,500 to 15,000 feet is called the oil window. The oil is less dense than the surrounding water, and the oil works upward toward the surface. Some of the oil, probably about 10 percent, gets caught in traps to form oil fields. The rest arrives at the surface as oil seeps or tar pits. Even the trapped oil can leak upward. If a billon-barrel oil field leaks one drop per second (a "drop" from a medicine dropper is about 1/20 milliliter), the leak will empty the oil field in 100 million years. The bottom line is that most oil is surprisingly young. For instance, in Wyoming the organic-rich source rocks and the porous reservoir rocks come in two flavors. In one set, the source rocks and the reservoir rocks are about 300 million years old. The younger set is roughly 100 million years old. If the source rock and reservoir are that old, isn't that the "age" of the oil? Nope, sorry 'bout that. The Wyoming basins continued to fill with additional sediment until the Early Miocene, 15 million years ago. Some of those older source rocks may not have been pushed down into the oil window until the Miocene. Youth wins. To a geologist, 15 million years ago is like yesterday.

Below 15,000 feet, further cracking reduces the oil to the simplest hydrocarbon, methane (CH_4). (There is an exception. Salt, ordinary table salt, conducts heat faster than the surrounding sediments. Some oil adjacent to salt layers goes a few thousand feet deeper than normal.) In addition, some organic matter—like wood and leaves—selectively produces mostly natural gas and little oil. And like oil, natural gas can leak to the surface.

Natural-gas seeps sometimes come out of harder rocks. At several locations, including Muktinath in Nepal, a crack in the rock supports a small natural-gas flame, as well as a flow of water. The superposition of fire and water made the village worthy of pilgrimages. (I made my pilgrimage to Muktinath in 1999.)

In softer rocks, natural gas seeps are usually expressed as mud volcanoes. They aren't volcanoes in the usual sense. Mud volcanoes are typically mud pits about twenty yards across, with occasional large gas bubbles breaking the surface. One mud volcano right on the coast of Venezuela has an additional attraction. You can take a mud bath for its reputed therapeutic value, then wade into the Caribbean and wash off the mud. Because there are mud volcanoes around the world, the famous petroleum geologist Hollis Hedberg had a favorite lecture titled "The Flatulent Earth."

The Ventura oil field, discovered in 1918, is still actively producing today. It is possible that a modest amount of additional oil has resulted from subsurface oil seeps during the lifetime of the field.

With all these surface oil seeps and natural-gas leakages, it comes as no surprise that some subsurface oil and gas fields might be actively receiving oil or gas. Unfortunately, the process is much too slow to make a difference in the world oil supply. The Ventura oil field, immediately north of the Los Angeles Basin, has been producing for more than a hundred years. Since there are surface oil seeps all around, it is to be expected that a tiny fraction of the Ventura oil would arrive after the field began producing. Similarly, oil and gas are actively migrating upward in the Texas and Louisiana Gulf Coast. However, the oil in each actively migrating area has the same chemistry, and molecular fossils, which serve to identify the source of the oil and gas as products of thermal cracking of organic-rich sediments.

Here's a warning: The conventional peak-oil story points to a grim and unhappy future. Combining an inorganic source for oil with active refilling of oil fields is an attractive alternative view. "Don't worry; oil and gas will last forever." A surprising number of people latch onto this false optimism. Further, it is an almost impossible task to get them to change their minds. My advice is to be patient with them; history will decide.

FOUR

The Dismal Science

F ifty-six years ago, I took a beginning course in economics. The textbook was Paul Samuelson's classic, but I didn't understand much. In fact, the only tidbit I remembered was that John Maynard Keynes married a glamorous Russian ballerina. Turns out, even that factoid was less than complete. In the ensuing years, the only economics book that I read and enjoyed was Burt Malkiel's *A Random Walk Down Wall Street.*

Today, when I see graphs that present economic data, the data have usually been massaged by correcting for inflation, converted to year-on-year changes, made into a running average, and based on a vertical scale somewhere above zero. It looks impressive, but I can't grasp the meaning. Most economic observations are ratios. The price of oil is given as dollars per barrel, but it could equally well be written as barrels per dollar. Almost daily, there is discussion of whether an increase in oil price reflects an oil shortage or a weakening dollar.

In an attempt to keep it simple enough for me to understand, the following graph shows the annual U.S. expenditure for crude oil as a percentage of the gross domestic product. No inflation correction, no averaging, no cooking the books, the bottom of the cost axis is zero. Surely, a zillion economists must have invented the same approach, but I'm too lazy to spend a dismal week in the library finding out who did it first.

The uncooked graph shows the United States' annual expenditure for oil (barrels consumed times the oil price) as a percentage of GDP in the same year. What jumps off the page are two spikes, one in the early 1980s and one in 2008. The two spikes have rather different origins, but their effects are similar.

The spike in the late 1970s is best understood by the story of the scorpion and the frog. A scorpion and a frog are at one side of a river and the scorpion wants to get to the other side. He asks the

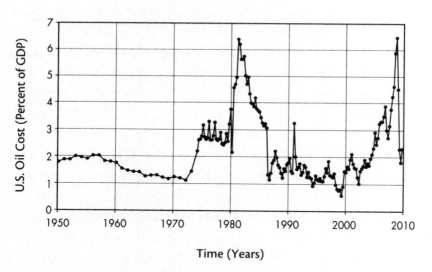

Monthly U.S. expenditures for crude oil as a percentage of gross domestic product. Expressing the oil cost as a percent of GDP avoids correcting for inflation. The 2008 oil cost spike is no higher than the 1981 spike, but the 2008 spike acted on a more fragile economy. The low price around 1998 brought us the SUV. (© Steve Deffeyes)

frog to cross the river carrying him on his back. The frog objects, "You might sting me." The scorpion explains that since he can't swim, he ought not to sting. Halfway across the river, the scorpion stings the frog. "Why did you do that? Now we both will die." The scorpion says, "Welcome to the Middle East."

The cause of the 1980s spike was conflict in the Middle East. The upward step around 1973 is the time of the Yom Kippur War. The Iran-Iraq War corresponds to a large price spike from 1980 to 1985. OPEC severely reduced oil exports, with painful repercussions in the United States. My point of reference for those years was one summer when it looked as if we could not afford a one-week family vacation. Eventually, we did scrape up enough money to drive to New Hampshire. The oil price rise rippled through the economy, causing double-digit inflation without expanding the economy. "Stagflation" was the new word.

After the 1980s spike, OPEC realized that the goose that laid their golden egg had been in intensive care. The goose eventually recovered, but the episode taught OPEC that they better not kill the goose. An OPEC strategy emerged for actively maintaining the oil price in the range of $22 to $28. Below $22, they would cut production. Above $28, OPEC would protect their goose by flooding the world with cheap oil.

Although the 1980s episode was painful, it did not have the crushing severity of the 1930s Great Depression. I have childhood memories of the Depression and the Dust Bowl, although I had no real understanding of the circumstances. I was transported back to that era when one of the first American Girl Dolls was Kit Kittredge, a child of the Great Depression. Right away, I bought a Kit Kittredge doll for my granddaughter. The doll's housedress could have come from my mother's closet. The doll came with several books, historical fiction, about Kit's cleverness in surviving the

1930s. If things get any worse today, we adults may be reading Kit's books for survival hints.

When scholarly histories are written later, there is an understandable desire to identify the triggering event. My favorite example is World War I, which started with the assassination of Archduke Ferdinand at Sarajevo. One museum even displays the bullet, with a caption: "The bullet that started World War I." Suppose that the assassin had missed. Would there still have been a World War I? Ya durn tootin'. Europe was in an incredibly unstable situation; some other event would have been the trigger. I'm tempted to rewrite the start of World War I, downplaying the importance of the assassination. I even have a working title: *Ferdinand: The Bull.*

For the present-day situation, conventional analysis blames the crisis on subprime residential mortgages. That's bull, as well. There were plenty of other instabilities in the U.S. financial situation:

- Large trade deficits, with overseas reinvestment back into the U.S. economy.
- A service economy, dependent on overseas sales of Coca-Cola, Hollywood movies, and Microsoft software. Other countries soon learned to make their own soft drinks, movies, and software.
- Overseas outsourcing of U.S. manufacturing capacity.
- Major dependence on imported oil.

In my view, the U.S. economy was like a house of cards, and it didn't matter which card got yanked out of the bottom of the house.

Unlike the 1980s oil crisis, in the present situation OPEC has little or no unused production capacity. World oil production ceased to grow as of 2005. (As early as 2001, I was predicting that 2005 would be the turning point.) Even in 2003, crude-oil prices started to rise, possibly because demand, particularly in China and India, was growing faster than world oil production was rising. The extreme price rises after 2005 were unable to coax expanded production out of the ground.

For years, I have been illustrating the consequences of an oil shortage with this slide:

I don't know how old the slide is, but it is getting yellow around the edges and brown in the corners. The most endangered industries are at the top of the alphabet; zymurgy is the use of yeast to make bread, cheese, and wine.

- Agriculture is at risk for several reasons. Manufacturing phosphate and nitrogen fertilizers is highly energy intensive. Transporting food to market uses diesel fuel. Flying up fresh fruits and vegetables from the southern hemisphere during the northern winter will become impossibly expensive.
- Big SUVs and pickup trucks are simply not selling. General Motors and Chrysler have gone through bankruptcy.

■ Half a dozen airlines have gone out of business. Most of the others are adding surcharges that they hope we won't notice. Airfreight, airframe manufacturers, and jet engine suppliers are equally at risk.

Is there anything that we can do right away?

■ A 50 mph speed limit (this won't go over big in Montana).
■ Carpooling (Internet-boosted commuting arrangements).
■ Opening and closing windows to minimize air-conditioning and heating.
■ Local agriculture. My family recently subscribed to a Community Supported Agriculture program. Delicious produce, and it hasn't been hauled across the country. Victory gardens from World War II are starting to appear in backyards.

There are lots of longer-term actions that we can take; many of them are discussed later in this book. However, it helps our attitude (and budgets) if we can do some things immediately.

FIVE

The Awl Bidness

T he oil business is 150 years old, it is huge, and its character is changing. Originally, the business was dominated by large private companies. However, the growing dominance of Rockefeller's Standard Oil Company caused the U.S. Supreme Court in 1911 to require Standard Oil to be broken up into "competing" pieces. Impressive pieces: Exxon, Mobil, Chevron, Amoco, Pennzoil, and others. In 1999, when Exxon (Standard Oil of New Jersey) was allowed to merge with Mobil (Standard Oil of New York), a distant alarm bell rang. Part of the 1911 decision was being reversed. (My father had to explain to me that the "Esso" trademark, still in use in Canada and Europe, was the phonetic spelling of the Standard Oil initials. A later one that I didn't get at first was the Q8 filling stations in England; they're the Kuwait Petroleum Company.)

These American companies, plus Shell and BP, had international operations, either on their own or in partnerships. Gradually, individual countries established their own oil companies or nationalized the existing partnerships. One landmark was the gradual conversion from 1973 to 1980 of Aramco (a partnership of Chevron, Texaco,

Exxon, and Mobil) into Saudi Aramco. The major international oil companies are gradually becoming engineering teams with a merchant bank attached. They have few opportunities to expand their wealth through the ownership of oil and gas discoveries.

Tar Sands

The last accessible frontier for liquid oil is the tar sands of Canada. Large tar-sand deposits exist along the Orinoco River in Venezuela, but access of international oil companies to Venezuela is progressively being reduced. The Canadian deposits are huge: a trillion barrels of oil, comparable in size to all of the conventional world crude oil produced since the beginning of the oil industry. Producing oil from the tar sands is not simply a matter of lapping it up.

The Canadian tar sands (which the Canadians prefer to call "heavy oil") are along the Athabaska River, mostly in Alberta but partially in Saskatchewan. There are two different extraction methods:

- Where the tar-sand layer is close to the surface, strip off the overlying sedimentary layer and mine the tar sand. Truck the tar sand to a separating plant. Heat the sand until the tar melts; separate the tar and haul the sand back to refill the excavation. In an on-site refinery, break the large tar molecules into smaller and more liquid molecules. Send the oil off by pipeline.
- Where the tar sand is deeper, drill holes through the overlying layers into the tar sand and use steam or solvents to make the oil fluid enough to be pumped out.

At this time, 87 percent of the Canadian tar-sand production comes from mines. In order to benefit from economies of scale, the mines

Satellite photograph of a tar-sand mining operation near Fort McMurray, Alberta. The disturbed area is about ten miles by ten miles. (Actually, I do not know which tar-sand mine is in this photo; I simply located Fort McMurray on maps.google.com and scrolled around looking for a large nearby mine.) (Courtesy of NASA Earth Observatory)

use giant scoops and enormous trucks to recover the tar sand. The capital cost for opening a tar-sand mine is more than $1 billion. But even if you have a few billion dollars burning a hole in your pocket, there are additional problems. Tar-sand operations are heavy users of natural gas, they use three barrels of water to recover one barrel of oil, and skilled construction workers and operating staff have to be brought in from outside. Companies are recruiting personnel from as far south as the Caribbean. Can you imagine someone from Jamaica facing his first 40-below-zero winter? Currently, tar-sand production from Alberta is close to one million barrels per day. Announced projects, including expansion of existing mines, would bring the production to about five million barrels per day.

The recovery of the deeper tar sands from drilled wells currently accounts for 17 percent of the Canadian tar-sand production. Using drilled wells is more like the conventional oil business, and the entry costs are mere millions instead of billions of dollars. Almost all the existing and proposed projects use steam injection to melt and mobilize the tar. A fascinating alternative to steam is the use of solvents to liquefy the tar, combined with later recovery of the solvent. Unfortunately, there is not an obvious choice of solvent. An ancient rule around the chemistry lab is "Like dissolves like." The dominant components in the tar are hydrocarbon rings linked together. The first solvent that comes to mind is benzene, but benzene is legally listed as a carcinogen. If you know of a good, cheap tar solvent that doesn't cause cancer, do not send me an e-mail. Head for Alberta and get rich, big Texas rich.

As Canada's internal needs for natural gas increase and conventional gas wells decline, alternatives for the tar sands are being explored. Some of the least valuable components of the tar could be burned to produce heat, or reacted with steam to produce hydrogen for upgrading the salable portions of the tar. A more radical solution, not now under construction, would be to build a nuclear reactor to produce hydrogen by electrolyzing water and use the waste heat to liquefy the tar.

National Oil Companies

In several episodes around the world, countries ejected the international oil companies and founded their own national oil companies. In some, but not all, countries the national oil company was viewed as a cash cow. In some countries today, oil sales produce half of the national budget. But the cow needs to be fed: Equipment has to be maintained, enhanced recovery projects are required, and new fields are needed to replace the declining production from existing fields. A government with newly nationalized oil production does not necessarily recognize the need to invest a portion of the cash flow to maintain future production. This is not unique to oil; rich mines in the copper belt across southern Africa gradually were closed through a lack of maintenance.

An important example of oil nationalization took place in Mexico. The anniversary of Mexican oil nationalization—March 18, 1938—is still celebrated as a national holiday. During the 1960s, I became acquainted with a geologist who had worked for Shell Oil in Mexico before the nationalization. He was a micropaleontologist,

A 1964 poster by Luis Arenal commemorating the 1938 expulsion of the international oil companies from Mexico. It was not what you would call a friendly takeover.

a specialist in the millimeter-size fossils that can be identified in the small rock chips from the drill bit. He had assembled a collection of these fossils, glued on microscope slides: the key to the subsurface history of Mexico. He was instructed not to take anything from the office. Here was his moral dilemma: Did his years of effort assembling the microfossil collection belong to Shell or to Mexico? He could quickly run his fingernail across the top of the slides, crush the tiny fossils, put the slides neatly back in their storage rack, and the collection would be valueless. If he got caught, would he spend the rest of his life in a Mexican jail? His decision was to crush the fossils. He did get out safely. Later, he was regarded as a corporate hero within the Shell organization.

Pemex, the national oil company of Mexico, went on to be a success. The culmination of their history was the discovery of the Cantarell oil field in the Gulf of Mexico. Cantarell was the world's largest field outside of the Persian Gulf; at its peak it produced more than a million barrels per day.

The main oil-producing horizon at Cantarell is described as "a breccia at the Cretaceous-Tertiary boundary." Breccia is a rock composed of sharp rock fragments and mineral cement. This is entangled with another story. Walter Alvarez in *T. Rex and the Crater of Doom* states that the dinosaurs were killed by a giant meteorite impact at Chicxulub, on the Yucatán coast, about seventy miles east of Cantarell. The mass extinction, killing more than the dinosaurs, is by definition the Cretaceous-Tertiary boundary. Was the breccia, the oil reservoir at Cantarell, caused by the meteorite impact? Or was the Cantarell breccia due to something more mundane, like dissolving a gypsum bed? Is the evidence for the Chicxulub meteorite impact a misinterpretation of a breccia layer? Stay tuned, this may take a while to resolve.

Despite strong efforts on the part of Pemex, including nitrogen injection, oil production at Cantarell is declining. Today, oil production for all of Mexico has dropped 16 percent from its peak in 2004. Similar production declines have happened in many countries; it is not a criticism of Pemex.

Pemex considers itself to be an oil company. Finding and producing natural gas is not their primary business. Each continent tends to be a single gas market; only a small percentage travels on liquefied-natural-gas ships between continents. For North America, NAFTA requires Canada to continue selling natural gas to the United States. The United States is obligated to continue selling natural gas to Mexico. It is only a raw guess on my part to say that the oil wealth of Mexico is probably accompanied by a wealth of natural gas. Discussions have taken place over natural-gas joint ventures between Pemex and the international oil companies. However, any international access is viewed locally as a sellout of the 1938 nationalization.

Around the world, there are countries that form joint ventures and countries that forbid outside participation. A joint venture between an international oil company and a local, national oil company is complex to negotiate and even more difficult to enforce. In only a few countries around the world can you write an enforceable contract. Here is an example: If the joint venture drills only dry holes, nobody cares. If the joint venture discovers a supergiant oil field, then the host country promptly suggests renegotiating the contract. "It's our oil, isn't it?" The bottom line is that the international oil companies, the "majors," have very few opportunities to increase their wealth by drilling in independent countries.

Another phenomenon is national oil companies that are owned 51 percent or more by the local government, with the rest of the common stock traded on stock markets. Examples are PetroChina and CNOOC in China and Petrobras in Brazil. At times, Petro-China exceeds ExxonMobil for the honor of having the world's largest market capitalization. PetroChina made a major discovery of light crude oil in shallow water offshore. I was delighted, because I owned a small block of PetroChina stock. If I understand it correctly, the government of China lowered the retail price of gasoline and diesel fuel and redirected a substantial portion of PetroChina's cash flow to benefit the Chinese consumer. Perfectly legitimate— remember that the government of China is the majority owner. I sold out my stock position, with some regret. My PetroChina stock was a souvenir purchase on a trip to Shanghai.

Petrobras, the Brazilian national oil company, has a structure similar to that of PetroChina. The government owns more than half of the stock; the rest is traded on stock exchanges. For the last few years, Petrobras has been reporting discovery after discovery in deep water offshore from Brazil. Petrobras exploration wells had an 83 percent success rate. Taken together, the offshore Brazilian wells were the largest oil discovery in twenty years. At the same time, there were a few published accounts about the possible formation of a wholly government-owned Brazilian oil company, with the offshore leases to be transferred to the new company. Uh-oh. I'm allergic to that sort of rumor; I sold out my Petrobras stock.

Holding Back Oil?

A 1974 paper by Robert Solow (Nobel Prize in economics) titled "The Economics of Resources or the Resources of Economics" pointed out that the owner of a mineral resource in the ground (including oil) could sometimes profit most by leaving it in the ground temporarily. The owner—whether an individual, a corporation, or a sovereign country—had to be wealthy enough to live comfortably without the cash flow. However, Solow was analyzing the case where all the players had access to full information. The owner of a rich zinc deposit would need to know the production costs and unmined reserves of other zinc mines as well as the price elasticity of zinc. For oil, oil field operators have little or no information about international oil reserves. It's a game they play blindfolded.

For several years, I have been watching for signs that a country or company was holding back oil. My first candidate was Abu Dhabi, part of the United Arab Emirates. They have lots of oil and a small population. About 90 percent of the residents are guest workers; there are only about four hundred thousand Abu Dhabians. The oil-production statistics for the UAE show a slow steady increase. I do not see a production decline that would indicate an abrupt decision by Abu Dhabi to hold oil in the ground. In contrast, on October 7, 2008, Shokri Ghanem, the chairman of Libya's national oil company, said, "With oil prices collapsing and international banks being routed, it's better to keep our oil in the ground."

The Majors

In the mid-twentieth century, the large integrated international oil companies were referred to as the Seven Sisters. Today, three of the seven have disappeared through mergers: Mobil, Gulf, and Texaco. Here is a list, as of July 2009, of the companies whose market capitalization is greater than $100 billion:

PetroChina	$374 billion	(China; majority government ownership)
ExxonMobil	$332 billion	(Standard Oil of New Jersey)
Petrobras	$151 billion	(Brazil; 51% government ownership)
Shell	$148 billion	(Netherlands)
BP	$143 billion	(formerly Anglo-Persian Oil Company)

Sinopec	$134 billion	(China; majority government ownership)
Total	$131 billion	(French; causes confusion in numerical tables)
Gazprom	$122 billion	(Russian; primarily natural gas)
Chevron	$121 billion	(Standard Oil of California)

Of course the ranking, and even inclusion in the list, changes daily on the stock market. Some additional large companies are not traded on stock markets. If they were, Saudi Aramco would probably appear at the top of the list.

Marketing is another matter. I keep seeing ghosts along the highway: filling stations named Getty, Citgo, Gulf, and Texaco. The original companies disappeared in mergers, but the successor company either kept the old name on the filling stations or sold the name to another company. Here are some examples:

Citgo	PdVSA, Venezuelan national oil company
Gulf	Cumberland Farms, operator of convenience stores
Getty	Lukoil (Russian company); some stations being rebranded as Lukoil
Arco	Merged with BP (formerly British Petroleum; formerly Anglo-Persian)
Shell	Joint venture, Shell and Saudi Aramco, especially in southeastern U.S.
Texaco	Separated from Chevron during 2001 merger; regained by Chevron in 2006

Before 1980, the return on invested capital was higher in the oil business than in any other industry. The major oil companies could not diversify into related businesses; everyone else was trying to diversify into oil. For the majors, the obvious choice was to invest a portion of the cash flow into improved ways of finding and producing oil. One important thing about industrial research: There is no second prize, no silver medal. In any subcategory (refinery catalysts, reflection seismic exploration, deepwater engineering) it is winner-take-all. Being able to bid confidently for deepwater oil leases is a tremendous competitive advantage.

After 1980, the major-company research budgets began to decrease slowly at first and then fall rapidly. The Hubbertian interpretation is that the companies realized, consciously or unconsciously, that searching for new oil would not be the wave of the future. Why pay to develop better exploration methods for an era when there is little oil left to find? Gradually, research leadership passed to the service companies.

The next activity to shrink was exploration. If there is not much oil left to find, should you be frantically drilling dry holes? The classic economic interpretation says that oil scarcity, and an associated price rise, would induce increased exploratory drilling. Another lesson from the late 1970s and early '80s: At any time, there are only a limited number of decent drilling targets ("prospects"). Drilling lower-quality targets almost always results in drilling dry holes.

I shuddered when some of the companies began buying back their own stock. They had their reasons, but to me a stock buyback was an admission that they had no meaningful way to invest the money in their own business. I developed a fantasy that the last employee of ExxonMobil would use the cash from the last barrel of oil to buy back the last share of stock.

Some of Hubbert's critics—call them neocornucopians—use the recent deepwater discoveries off the U.S. Gulf Coast and off Brazil as evidence that major new finds could rescue us from a global oil shortage. There are conflicting pieces of news:

Rock salt (ordinary table salt, sodium chloride) can flow slowly under pressure. Near the center of this photograph of Iran (taken from a spacecraft) are two black salt glaciers extruded from anticlines (rock domes). The salt itself is white, but the few rains in this arid climate dissolve away some of the surface salt, leaving black minerals as a residue. If you would like to be a geologist, notice the twin light-colored layers that outline the anticlines. (Courtesy of NASA Earth Observatory)

- Imaging geologic structures beneath a salt layer was a major challenge. One of the credits after the Jack 2 discovery in the Gulf of Mexico was for the mathematics that imaged the structure. It's a great day when the nerds are given part of the credit.
- The April 2010 blowout of the BP well in the Gulf of Mexico forced delays, and even cancellation, of deep-water drilling programs.

The salt layer is sodium chloride, ordinary table salt, deposited from the evaporation of ancient seawater. There are two kinds of subsalt oil fields: "presalt" production from layers deposited before the salt layer was deposited, and "subsalt," where the original salt layer flowed upward and mushroomed out. (Salt flowage is not limited to subsea areas; there are surface salt glaciers in Iran.) The Gulf Coast discoveries are subsalt; the Brazilian finds are presalt.

In addition to the ability to flow, salt is a good conductor of heat. As a result, temperatures are somewhat lower in the presence of salt layers and salt domes. In the absence of salt, most sedimentary basins have the bottom of the "oil window" at about 15,000 feet. Some of the new discoveries find oil 17,000 feet beneath the sea floor. One warning: Discovery reports usually give the well depth and water depth, but it does not automatically follow that the oil find is at the very bottom of the hole. A wonderful side effect of being near the bottom of the oil window is light crude oil. One of the recent Brazilian discoveries isn't quite gasoline, but it is as light as diesel fuel or jet fuel. Fill her up.

The U.S. Gulf Coast subsalt discoveries consist of several widely separated wells. The size of the new resource is not yet clear. Off Brazil, the size gets bigger with each new exploratory well. In oil field slang, "They're cutting up a fat hog." Does this solve our global oil problem? Unfortunately, when the Brazilian discoveries started, there were only a dozen drilling rigs big enough to deal with the water depth and the rock depth. The Brazilian national oil company, Petrobras, began signing contracts for the future services of three-quarters of the world's big deepwater rigs and they sponsored construction of additional rigs. Good move, but now the possible political movement in Brazil to reassign the offshore leases could leave Petrobras as little more than a drilling contractor. The bottom line is that Brazil will help the world situation, and the first subsalt oil is now being refined. However, the time constant for getting substantial new production on line is ten years, and getting longer.

New Technology

A standard "wish" is for new technologies that radically improve the economics of oil production. All along, the profitability of a successful oil company was sufficient incentive to find more effective ways of doing business. However, most of the "improvements" have been incremental changes in the existing technology.

A recent example is the long core barrel developed by Baker Hughes. For sixty years, solid rock cores were cut from wells with a core barrel either 60 or 90 feet long. Every 60 or 90 feet, it was necessary to pull the drill pipe out of the well to retrieve the core. The new Baker Hughes gadget has taken single cores as long as 622 feet, often enough to cover the entire productive section. As a geologist, I'm delighted. A three-inch-diameter core yields information that is not available from well logs and drill cuttings.

Should we be looking for radical new technologies? The equivalent of Google? In a limited sense, the coil-tubing drilling rig is the first fundamental change in one hundred years. So far, coil-tubing rigs have not dominated the industry, but they are gaining market share.

If we were to try, who would pay for the research and development? Historically, it would have been the research labs of the major oil companies. Company research labs have been severely reduced in scale. The larger service companies are now the leaders in innovation, but they are not likely to take on radical new schemes. Statistically, most of these programs fail.

Speculators, Like Me

As oil prices shot upward in 2008, there was a widespread search for a villain. Oddly, M. King Hubbert was seldom cited as the villain in chief. Here's a list:

- The major oil companies were swimming in cash, but they insisted that they lacked access to good places to drill.
- OPEC was not behaving like a good cartel behaves; even with all the valves wide open, OPEC could not hold down the oil price.
- Governments in oil-importing countries claimed they didn't see it coming. Both *Hubbert's Peak* and *Beyond Oil* are available at amazon.com for less than $15 each.
- Speculators, financial firms, and particularly hedge funds did see it coming and exaggerated the price swings. Isn't that what capitalists are supposed to do?

I hereby confess that I implemented my own little speculative scheme. For oil companies, divide the enterprise value (market cap minus debt) by the company oil reserves. Sort the resulting list in dollars per barrel of reserves, lowest price at the top of the list. Go down the list scratching off companies with less than a one-billion-dollar market cap (too volatile), with large short positions (somebody might suspect something), or with recent price drops (bad management). The basis of my scheme was buying cheap oil in the ground, knowing that peak oil would raise the price. Like so many schemes, it worked fantastically well as long as the market for oil stocks was rising. I wasn't selling short, but I was buying more oil than I intended to consume. The price crash in the second half of 2008 caused me to sell out my speculative investment. I'm hiding in gold for the moment, but I hope that it is temporary.

SIX

The Wider Picture

L et's zoom out, put on the wide-angle lens. We can best understand the energy crisis in the larger context of the world's resources. I explicitly reject a people-are-evil methodology. People are frequently misguided or ill informed, but only a few of us deliberately choose evil behavior.

There are a large number of options and some inherent constraints. Only a tiny amount of platinum is available to build fuel-cell cars. Identifying the constraints is an important first step in choosing our present-day moves. We don't want to redesign our economy to run on natural gas and then run short of natural gas. Although I will try to identify some of these constraints, be warned that I probably will not identify all of them. In that sense, this is a methodology model. It needs to be expanded to lower the probability that an important constraint will be overlooked.

The first resources-constraint study to gain wide public attention was the 1972 Club of Rome analysis *Limits to Growth*. It was a disaster. At the core of the study was a computer model of the interactions in the global economy. In their eagerness to get to the fun

part—running the model—the authors tossed in some quick estimates of the constraints. (In 1972, there was no such thing as
Google. Google's founders, Sergey Brin and Larry Page, were born
in 1973.) Their computer model predicted a resounding crash in the
world economy around the year 2000. Not only did the world survive 2000, the false alarm rubbed off on all other predictions of
future constraints.

In particular, the Club of Rome included no geologic information in setting their limits. To me, of course, omitting geology is a
mortal sin. Copper, agricultural soil, iron, groundwater, and oil are
geological entities. Usually, the question is "How much is available,
at what price?" There isn't a simple, fixed lump of copper. If you are
willing to process leaner metal ores, grow crops in marginal soil, or
drill deep natural-gas wells, then the size of the resource increases.
So let's start a new entity to look at resource availability. We'll name
it the Club of Wasilla.

The Matterhorn Project

A frequently heard response to the oil crisis is "Why don't we start
another Manhattan Project or Apollo Project?" The Manhattan
nuclear weapon and the Apollo lunar landing had goals identified at
the beginning. (I was in Rice Stadium when President Kennedy
announced the Apollo Project; it was heady stuff.) For the oil crisis,
the nominations from the floor include:

- Drill, baby, drill (more oil)
- Ethanol from corn, sugarcane, or carbohydrates
- Clean coal gasification
- Unconventional natural-gas reservoirs
- Solar, both photovoltaic and heat

- Oil from marine algae
- Geothermal energy
- Tidal and wave energy
- Hydroelectric
- Oil shale
- Nuclear reactors

Do we need ten or twenty superprojects? What if we place our bets on one and it doesn't work out? Let's look at the Matterhorn Project.

In the early 1950s, the Princeton astronomer Lyman Spitzer Jr. was on a family skiing vacation in Aspen. While riding up the ski lift, Spitzer had the insight that he could mimic the thermonuclear reactions inside the sun. A circular vacuum tube surrounded by magnetic coils might produce an energy source with no radioactive waste. Spitzer's wife reported that it was the worst family vacation they ever had. Lyman had his great idea on the first morning of the vacation. He got off the lift at Midway, skied down, and spent the rest of the week holed up in the hotel room, writing.

Nuclear energy can come from either end of the scale. The lowest energy, most stable configuration, is in the middle with iron or nickel. At the heavy end, you can fission uranium nuclei into pieces and get energy out. At the light end, you can fuse together hydrogen nuclei and get out even larger amounts of energy. Our sun is just a big thermonuclear fusion reactor in the sky.

Work on Spitzer's controlled fusion reactors began promptly, but initially it was a secret, classified project. (In the early 1950s, hydrogen-fusion thermonuclear weapons were under development.) The work at Princeton, known as the Matterhorn Project, was declassified in 1958. I happened to be at Princeton as a graduate student and I eagerly showed up for the first tour of the facilities. I've been following the project since, sort of cheering from the sidelines. Lately, it hasn't been looking good for the home team. Over the fifty years at least $10 billion have been spent around the world to develop practical fusion energy, and we still are quite a ways from having a practical system installed. The original World War II Manhattan fission-bomb project cost "only" $3 billion. The lesson for today, friends, is that these massive projects do not necessarily deliver the product on time and under budget.

The Solar Future

The geological good news is that we have deposits of energy sources to last us for at least one hundred years. The environmental bad news is that the deposits are of coal and uranium. After coal and uranium are no longer available, our civilization will have to be 100 percent directly or indirectly solar: biomass, wind, hydroelectric, and waves. Make that 99 percent; tidal energy and geothermal heat are not solar. There is an important message here. On our way to our solar future, we do not want to bet on a loser. I mentioned natural gas as an example earlier. There has been good news lately about recovering natural gas from organic-rich shales. We don't want to rebuild our economy on natural gas and have it crash before we are fully solar.

Although sunshine and wind are distributed globally, they are not distributed equally; there are hot spots and fairways similar to the fossil fuels. As an example, southeastern Wyoming is the wind capital of the United States. At the moment, developing additional electrical wind generation in Wyoming is limited by the capacity of the existing electric transmission lines.

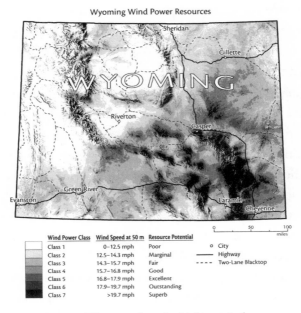

Wyoming Wind Power Resources

Wind Power Class	Wind Speed at 50 m	Resource Potential		
Class 1	0–12.5 mph	Poor	o	City
Class 2	12.5–14.3 mph	Marginal	——	Highway
Class 3	14.3–15.7 mph	Fair	- - - -	Two-Lane Blacktop
Class 4	15.7–16.8 mph	Good		
Class 5	16.8–17.9 mph	Excellent		
Class 6	17.9–19.7 mph	Outstanding		
Class 7	>19.7 mph	Superb		

Southeastern Wyoming has the highest wind velocities for electric power generation. The highest average wind velocities are mostly in the mountains, but there are some high plains sites with strong potential for power generation; an example is about fifty miles north of Cheyenne. I can testify personally to the wind intensity; I went to high school in Casper. I was there during the Great Blizzard of '49. (© Steve Deffeyes)

King Coal

Coal was the initial driver of the Industrial Revolution. Starting in the late 1700s, the revolution gradually blossomed into steam engines, railroads, ships, and steel mills. The great industrial economies initially grew where there was access to coal and iron ore. The United States and Northern Europe became the "first world." What later became the Soviet Union was called the "second world," and people wondered why the "third world" didn't have much coal. It was built into the Industrial Revolution. The rise of the oil industry diminished the need for coal for transportation, but coal continued to be used for generating electricity and producing metals.

Burning coal produces a host of environmental problems. I will point out later that coal gasification may be a way out of the problems, but first let's look at the environmental impacts.

More than half the weight of coal is carbon. When the coal is burned, the carbon goes up the smokestack as carbon dioxide. There is no question that burning fossil fuels—coal, natural gas, and oil—has increased the carbon dioxide content in Earth's atmosphere. (I'll consider the climate implications later.) From an environmental standpoint, conventional coal is the dirtiest fossil fuel.

Carbon dioxide, big as it is, is not the only environmental impact of burning coal. Coal typically contains a few percent of the mineral pyrite (iron disulfide, fool's gold, FeS_2). When the coal is burned, the sulfur converts to sulfur dioxide. Visit any remote Scottish village in cold weather; houses are heated with coal and outside you can smell, and taste, the sulfur dioxide. In the atmosphere, oxygen and water combine with the sulfur dioxide to produce sulfuric acid. That's acid rain; that's the killer smog in Donora, Pennsylvania (1948), and in London (1952). The acid rain is not limited to the immediate surroundings of the coal-burning plant. Acid rain falls out hundreds of miles downwind from the source. New England lakes and ponds suffer from acid sulfur pollution coming from the industrial fairway between Pittsburgh and Chicago.

Of the trace metals in coal, mercury has attracted most of the attention. The toxicity of mercury has long been known. Examples are the Mad Hatter in *Alice in Wonderland*, the mules in the Mexican patio process for recovering silver, and the workers in Nevada silver extraction mills, 1860 to 1900. Modern concern was increased in 1956 by methyl mercury, converted from mercury pollution out of a chemical plant, in Minimata, Japan. In 1997, a final message came when a distinguished Dartmouth chemistry professor died after spilling a few drops of dimethyl mercury on her rubber glove. The bottom line is that liquid mercury is not good for you but it does not kill right away: Methyl mercury causes slow brain rot; dimethyl mercury is endsville. Notice that these mercury horror stories are not from coal.

Is there any way out? Rejiggering existing coal-fired electric plants is tough. You have a big smokestack full of hot gas, and the competitive price of electricity limits the treatment options. It's one of those examples where it makes sense to start over from scratch instead of trying to retrebel the wheezetiller.

The most promising alternative for coal dates back to the late 1800s. Coal was burned with a limited amount of air and lots of steam. One option in the process was alternating sending in a blast of air (and producing mostly carbon dioxide) and sending in steam (and saving the hydrogen and carbon monoxide). The hydrogen and carbon monoxide gases were sold for lighting and as heating fuel. The increasing availability of natural gas gradually shut down most of the coal gas business.

There were two episodes in the mid-1900s when a major economy was cut off from supplies of oil: Germany during World War II and South Africa during the apartheid era. In both instances coal gasification was resurrected and the carbon monoxide and hydrogen were turned into gasoline by using an appropriate catalyst. The companies that did the work are still in business: Lurgi in Germany (now a division of Air Liquide) and Sasol in South Africa.

Engineers at Shell and Texaco improved coal gasification by using pure oxygen, higher temperatures, and better catalysts. Plenty of opportunities exist for using the gas, today known as synthesis gas. The gas can be burned directly to generate electricity and/or to provide heat. Hydrogen is used in turning atmospheric nitrogen into fertilizer and in petroleum refining (hydrocracking). Besides gasoline, diesel, and jet fuel, other products are methanol, ethanol, and dimethyl ether. The original Sasol plant in South Africa is now producing petrochemical feedstocks.

As mentioned earlier, burning coal in a conventional electric power plant sends several undesirable materials up the smokestack. Coal gasification initially produces a much smaller volume of gas, the gas is inside the plumbing, and unwanted materials can be removed.

In a coal gasification facility, the carbon dioxide can be separated. At the moment, there is a large market for carbon dioxide; it isn't Coca-Cola. Carbon dioxide can enhance oil recovery in partially depleted oil reservoirs. Eventually, carbon dioxide will have to be buried, possibly in the geologic subsurface.

Sulfur, another unwanted smokestack component, can be separated from coal gas. The technology exists for removing sulfur from natural gas. The sulfur can be recovered and sold. As with the carbon dioxide, it's great to be getting paid for your former pollutants.

Mercury is present in small amounts in coal; recovering it won't help the cash flow. Mercury is not only toxic but it gets increasingly concentrated as it is passed up through the natural food chain. Large predatory fish are an example. People who eat lots of large predatory fish are an even better example. (I love swordfish and tuna, but I'm cutting back to once per month.) The preferred existing mercury-removal technology uses steam-activated charcoal. Hard coconut shells make a particularly effective charcoal. An alternative, but overly expensive, technology would use silver or gold. Metallic mercury and silver (or gold) have an enormous attraction for one another. Probably most readers of this book have dental fillings made from a mercury-silver amalgam.

If coal gasification is so wonderful, why aren't we using it? We are, but on a small scale. Lots of entrepreneurs have written business plans for coal gasification plants. You want a coal layer, which can be low-grade coal. You need some oil fields within a few hundred miles to buy your carbon dioxide. Fresh water, cheap land, and a lax regulatory agency will help. The biggest barrier is uncertain future energy prices. The entrepreneur hates uncertainty; the coal gasification plant has to return the large initial capital expenditure. A second barrier is the availability of capital. Borrowing the money or raising it through a public stock offering is currently almost impossible. Help from the federal government isn't likely; there are lots of other eager hands reaching out for governmental aid. Coal gasification belongs on the short list of options being considered. If it is chosen as a major component of our energy future, then it will take strong political leadership to make it happen.

SEVEN

Uranium—the Red and the Black

I t's an exaggeration, but the legend claims that, from 1950 to 1980, there was a radioactivity counter in every pickup in the American West. Everyone was looking for uranium. The end of the era was caused by a reduced need for power plants and military use and the fact that we had found plenty of uranium.

During the 2008 presidential campaign, Senator McCain expressed his strong enthusiasm for building forty-five new nuclear-power plants in the United States. The Chernobyl disaster happened in 1986. The most recent U.S. nuclear-power plant was completed in 1996. The nuclear waste disposal site at Yucca Mountain doesn't look effective or safe. On the upside, nuclear-power plants do not add carbon dioxide to the atmosphere and we do not have to wait for an R&D program in order to build reactors. An enormous tug-of-war could ensue.

The preferred designs for nuclear-power plants are heavily entangled with the future uranium supply. From the beginning of the nuclear age, the capital cost for building the reactor loomed far larger than the cost of the uranium fuel. A 1970 nuclear-power plant was typically a "uranium guzzler," the nuclear equivalent of a 1970 Cadillac. The late-1970s discoveries of several large and high-grade uranium deposits in Canada and Australia indicated that the uranium guzzler could be around for a while. (I am aware of the heated debate about whether the government "subsidizes" the nuclear electric-power industry. Part of the debate centers around whether indirect benefits from nuclear weapons development should be counted as subsidies.)

Before the impact of the 1970s uranium discoveries was felt, there was concern that nuclear generation of electricity could be strangled by the limited uranium supply. The power in conventional nuclear reactors comes from uranium 235, which makes up only 0.7 percent of natural uranium. However, uranium 238, which makes up most of the remaining 99.3 percent, can be used indirectly. In an ordinary nuclear-power reactor, some of the neutrons from fissioning (splitting) uranium 235 get captured by uranium 238 to produce plutonium. The separation of plutonium, called "reprocessing," has to be carried out in a heavily shielded remote-control facility because of the intense radioactivity. Spent fuel rods from a few power reactors have been reprocessed, mostly in England. Having plutonium available as a commercial item raises serious concerns: it is highly toxic and it has to be protected from terrorist attacks. John McPhee's book *The Curve of Binding Energy* contains extensive interviews with Ted Taylor, who was America's most successful designer of nuclear weapons. The previous wisdom was that building a nuclear weapon involved an enormous industrial complex; you couldn't build the bomb in your garage. In McPhee's book, it becomes clear that Taylor *could* build a nuclear bomb in his garage if he were given suitably enriched uranium. When the subject turned to plutonium, Taylor refused to discuss it. The apparent conclusion is that almost any physics graduate student—given the plutonium— could build a nuclear explosive in his bathroom.

During the 1970s, engineers and economists working for the Atomic Energy Commission (merged into the Department of Energy in 1977) produced a report saying that uranium deposits would become scarce and the United States needed to begin building breeder reactors. A breeder reactor is designed to turn its uranium-235 input into a larger plutonium output. In effect, the breeder economy repeatedly recycles plutonium as a catalyst to burn the more common uranium 238. Breeders are very different designs as compared to conventional uranium reactors. Breeders frighten me. In a smaller way, I was frightened because the AEC-DOE report contained *no* geology. It was a replay of *Limits to Growth*. Who needs geology? We'll just plug some mathematical curves into the computer and if the curve of ore grade versus uranium recovery has the right shape, we build a national network of breeder reactors. Double panic.

I received a grant from the AEC-DOE to work through their archive of uranium-ore purchases to develop a geological curve of ore grade versus uranium recovery. The narrative of my adventure is told in *Beyond Oil*. The result of the study showed that each time you cut the uranium ore grade in half, you opened up about eight times more total uranium. (My fellow nerds don't have to be told that using a logarithmic scale for the vertical axis converts a Gaussian bell curve into a parabola opening downward.) The results of the study were published in the January 1980 issue of *Scientific American*. There was—and is—no justification for a breeder reactor. The operating commercial American breeder reactor (at Fort St. Vrain, Colorado) and the ambitious French *Phenix* and *Super Phenix* breeder reactors were eventually closed down.

As a by-product of that uranium experience, I developed a long-term interest in the origins of uranium deposits. That interest expanded into other ore deposits with similar origins. Historically, most ore deposits were thought to be deposited by hot-spring water

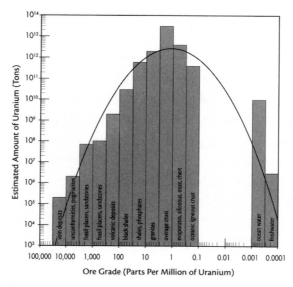

The slope of the line in the lower left-hand corner shows that cutting the uranium ore grade in half produces more uranium by a factor of eight. Therefore, there is no need for breeder reactors. The logarithmic scales on both axes reflect what is known as a "lognormal" distribution, seen in everything from sand grains on the beach to the abundance of lead in granite. (© Steve Deffeyes)

on its way to the surface, and that explanation accounts for more than half of all known ore deposits. The hot-spring interpretation was originally developed by the Swedish-American geologist Waldemar Lindgren, based on his observations in Nevada. Lindgren and his followers classified ore deposits as hypothermal (deep, 400–600°C), mesothermal (200–400°C), and epithermal (close to the boiling temperature of water, 100°C). Gradually, a small category of ore deposits came to be recognized as telethermal—not thermal at all, room temperature. Among the telethermal deposits were the uranium deposits on the Colorado Plateau (Utah, Arizona, New Mexico, and Colorado).

A few of the Colorado Plateau deposits were mined for vanadium between 1900 and 1940. Uranium was also present in the deposits, but there was almost no market for uranium at the time. (Vanadium was used to alloy steel.) As uranium exploration expanded after 1945, one subset of the deposits in sandstones gained a lot of attention. These were called roll-front deposits and they came to be recognized as a boundary between normally oxygenated groundwater and the same water whose oxygen had been used up to oxidize bits of organic matter in the sandstone. Dozens of these roll-front deposits were discovered on the Colorado Plateau and in Wyoming and Texas. Gradually, a weird pattern emerged: Roll fronts seemed to be an American specialty. A few roll-front deposits have been found in central Europe and in Kazakhstan. It still makes me nervous; everyone in the industry knows the roll-front story and I have no reason to think that the western United States is unique.

Sometimes a piece of knowledge in one field expands into an adjacent realm that initially did not seem to be related. Even without a big research and development program, we can take the idea and run. An example in chapter 9 was expanding the production of natural gas from the Barnett Shale into a worldwide new source of natural gas.

Red bed Copper

The chemistry that generates roll-front uranium deposits can occur for other metals and in other geologic settings. In the next three pages, I am going to focus on copper. The price of copper has tripled since 2003; China in particular is buying copper because of their need to expand their internal capacity for distributing electricity. However, the eventual expansion of the concept is not limited to copper. In addition to copper, other metals can be concentrated by the roll-front chemistry.

- It was not just uranium and vanadium. Numerous chemical elements change their solubility in water as the water changes from oxidizing to reducing. ("Reducing" is a chemical term for substances that can consume oxygen.) In the list of oxygen-sensitive elements are arsenic, cadmium, cobalt, copper, gold, iron, lead, manganese, molybdenum, nickel, selenium, uranium, vanadium, and zinc. In addition, several other chemical elements do not themselves participate in oxidation reactions, but other oxygen-sensitive elements (such as sulfur) can precipitate them out of solution, such as silver and barium.
- It isn't just sandstones with flowing groundwater. Oxidation-reduction boundaries occur wherever black organic-rich shales (including mature oil source rocks) are in contact with sandstones or permeable limestones. When I visited active open-pit gold mines in Nevada, I always asked to see the good stuff, the richest gold ore. In all three instances (Alligator Ridge, Jarrett Canyon, and Gold Bar) I was shown soft, black, sooty rock.

The most studied example of an oxidation-reduction ore deposit is the Kupferschiefer in Germany and Poland. The Kupferschiefer (with "copper" in its name) produces modest amounts of newly mined copper each year; its fame results from having a significant population of very capable geologists in the region. Although Stendhal already copped *The Red and the Black* as a title, it applies to the Kupferschiefer. The red is a sandstone, the black is an organic-rich shale. (I have to be careful here, because the area is the world standard section that defines the Triassic.) At the contact between the red and the black is the Rotliegendes, the "red rotten rock." It's rotten and red because it is black rock that has been oxidized. However, the most exciting feature to me is the sequence of minerals, starting in the oxidized red sandstone and proceeding through the Rotliegendes to the reduced black shale:

- Pyrite, FeS_2 (fool's gold)
- Chalcopyrite, $CuFeS_2$
- Bornite, Cu_5FeS_4
- Covellite, CuS
- Chalcocite, Cu_2S

The exciting part is the sequence. Except for pyrite, the sulfur is always S^{--}. In pyrite, one of the two sulfur atoms has to be oxidized to free sulfur, S^0. Copper stays as Cu^{++} until, at the reduced end, copper is present as Cu^+. The underlying hint is that the Kupferschiefer is an active system. The copper wasn't dumped in there millions of years ago and allowed to equilibrate with the neighbors.

The generic name for these deposits is red bed coppers. The biggest of them, the Copperbelt of southern Africa, is unfortunately only partially studied. The early reports show the same sequence of copper minerals given above. However, the early interpretation placed the pyrite-chalcopyrite end as a shallow-water oxidized deposit and the chalcocite as a deep-water oxygen-deficient sediment. The few recent reports mention organic-rich black shale interbedded with the copper ores.

At one time, the African Copperbelt was listed as running from Zaire into Zambia and then into Zimbabwe. In 1997, Zaire broke up my ZZZ mnemonic by changing its name to Democratic Republic of the Congo. Potentially, these are the lowest-cost copper mines in the world, but at the moment they contribute only 4 percent of the world's copper production. A huge capital investment would be required to rejuvenate the mines. As a small bonus, I would get new detailed descriptions of the copper deposits.

A side benefit from the oxidation-reduction story is focusing exploration for new deposits. Instead of trying to explore the entire area, the exploration geologist can track along the line of the red-black interface. I tried it once, but it was only a tiny attempt. There were several historical copper mines in the red beds of the New Jersey Triassic. (The upper third of the New Jersey section is actually Jurassic in age, but we're trying not to be picky.) A town named Copper Hill is about fifteen miles from Princeton. When I was teaching at Princeton, I explained the red-black story to my sedimentology class and we scheduled a one-day excursion to an area with previously mapped black shale units among the red beds. We set out to take samples along the red-black interface. It rained like crazy, and we got only two or three samples, none of them proving to contain anomalous amounts of copper. At least my students got a lesson in the difficulties surrounding mineral exploration. I'm still interested. Somewhere out there, there ought to be a red bed silver deposit.

So how do you go about it? Here are some hints:

- It isn't just the red and the black, although you could name your enterprise "Stendhal Exploration Company." All oxidation-reduction systems are in play.
- Historic mines, even if small, contain hints about processes that actually happened.
- Oil and natural gas migrating through the system are powerful reducing agents. In particular, the puzzles in the Mississippi Valley lead-zinc deposits, if unraveled, could be powerful clues to guide further exploration.
- In low-temperature deposits, don't overlook the role of bacteria.

Mineral exploration is not a first choice career for most students today. There have been only modest research investments by the mining companies, so exploration for new deposits is open to innovative ideas. Outdoor work, overseas travel, geology and chemistry; I regret that I'm too old to start a new career.

Radwastes

Longtime storage of radioactive wastes is widely cited as a barrier to further building of nuclear reactors. Storing radioactive wastes is an easy problem, turned into a hard problem for political reasons. If you are opposed to nuclear reactors because of your fear of another Three Mile Island, another Chernobyl, then you first set the standards very high: absolutely positive containment for ten times the longest radioactive half-life in the waste. Then you use NIMBY (not in my backyard) to make the good disposal sites inaccessible. Finally, you attack the remaining site (Yucca Mountain, in nobody's backyard) as not meeting the standards.

The reliable radwaste disposal sites are in geological layers of salt or calcium sulfate (the mineral anhydrite). These two rocks are the tightest seals that keep oil and natural gas deposits from leaking to the surface. (Again, the details are in *Beyond Oil*.) Beds of salt and anhydrite are available in several different states. The active Waste Isolation Pilot Plant (WIPP) for disposing of military radioactive wastes is in a salt bed east of Carlsbad, New Mexico. WIPP is accepting waste, although it may also be attracting extraterrestrials to nearby Roswell.

The United States will probably get over its distaste for nuclear power and will choose to build additional commercial reactors. The public may be ready to accept the traditional water-cooled reactors. The hottest (literally and figuratively) new reactor is the pebble bed reactor. The uranium (or plutonium) fuel is cast into baseball-size "pebbles," then coated with graphite. The heat from the reactor is

The Waste Isolation Pilot Plant, east of Carlsbad, New Mexico, is excavated in an underground salt layer. A mining machine is being used to cut a cubbyhole to house a canister of radioactive waste. I have been down in the Carlsbad salt mines; they really are as dry as it looks in this photograph. (Dirk Roberson, WIPP photographer)

removed, and utilized, by flow of a gas, preferably helium. Part of the attraction is stability. If a pebble bed reactor starts to run away and heat up, the rising temperature inherently slows down the reactor.

One final way of sugarcoating the renewal of nuclear-power reactors would be burning up part, or all, of the inventory of military nuclear weapons. Nuclear weapon inventories are being reduced in Russia and in the United States. Using the fissile ingredients from bombs to generate electric power is the modern equivalent of "they shall beat their swords into plowshares."

EIGHT

Climate Change

In 2006, I attended a meeting in Europe for authors who had written on peak oil or on global warming. The purpose of the meeting was for the two groups to understand one another. The meeting was a disaster; the two sides left even farther apart. The peak-oil writers were data driven, lots of numbers. The global-warming group seemed (at least to this biased observer) to be making an emotional case. They actually said that if I didn't agree with their story, then I must be suffering from some sort of mental disease or defect. They volunteered to treat my mental condition so that I would agree with them.

On the trip back across the Atlantic, I happened to share space with one of the climate-change writers who lived on the outskirts of Washington, D.C. He explained that he had searched out a source for environmentally acceptable dry corn. He located a Mennonite farmer who used horse-drawn equipment and organic fertilizer. He verified that the farmer's seed corn was organic and not genetically modified. So what did the global-warming writer do with the corn? Corn bread? Polenta? Tortillas? No, he burned the corn kernels to heat his house. Out in Oklahoma, we knew that corncobs would burn and we made cornbread out of the kernels. I did not volunteer to cure his mental condition; he seemed to be enjoying it.

One of my scientist friends, renowned for his deep intuition, pointed out that global warming is actually an engineering problem. You have a system that you want to understand, you need to select the output that you want from the system, and you need to find the least-cost way for getting the system to give the desired output. Hard work but very little emotion. So let's appoint ourselves to the engineering staff and have a look at the task.

Starting from Scratch

The Keeling graph for atmospheric carbon dioxide is a good place to begin. Since 1958, measurements have been made atop the Mauna Kea volcano in Hawaii. There is no question that carbon dioxide levels have been increasing, and mass-balance calculations show that burning fossil fuels is the major contributor to the increase.

The next engineering task is to ask about the effects of the added carbon dioxide. Fifty years ago, carbon dioxide was identified as an atmospheric component that traps outgoing infrared radiation. Holding back that infrared warms Earth.

Atmospheric models used for short-term weather predictions make useful forecasts for about five days ahead. (The forecast for the sixth day is usually "partly cloudy.") These models can be run to investigate long-term climate trends. The climate model is used as a what-if tool. If the atmospheric carbon dioxide is increased in the model, how do the temperature, rainfall, and winds change for each region? There are two major difficulties with the model:

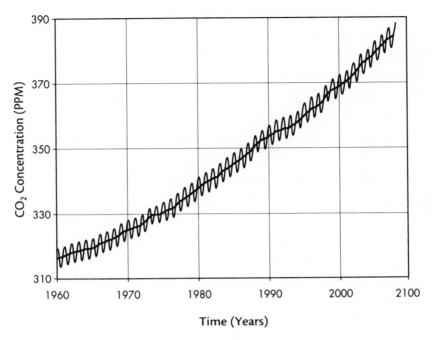

The buildup of carbon dioxide in Earth's atmosphere has been
measured since 1958 near the summit of Mauna Kea in Hawaii. The
measurements were made by Charles D. Keeling (1928–2005) of the
Scripps Institution of Oceanography; the measurements have been
continued by his son. The annual zigzags in the data are caused by
seasonal uptake of carbon dioxide during the Northern Hemisphere
summer. Although the data seem enormously valuable today, during
part of the time Keeling had difficulty keeping the measurement
program financed. (© Steve Deffeyes)

- The computer model runs on a grid of points in the atmosphere.
  On scales larger than the grid spacing, the underlying physics
  can compute the interactions. On scales smaller than the grid
  spacing, the behavior has to be "parameterized." Statements
  about the average behavior on small scales are embedded in the
  program. As computers got faster and memories got larger,
  the grid sizes became smaller. More physics, less dependence
  on averaged parameters. The latest enthusiasm is pulling the

processor and graphics chips out of game-box computers to make huge, but inexpensive, computer arrays.

- Earth's atmosphere responds on a time scale of days. The ocean takes one thousand years to exchange deep water with shallow water. Even the largest, fastest superdupercomputer is not going to run the model to examine thousands of years. Yet the oceans, and particularly sea-surface temperatures, have a major effect on the weather and climate. Some clever tricks have been used to explore the interactions between the two time scales, but the results still leave me a bit unsatisfied.

The first use of atmospheric models to examine the effect of adding carbon dioxide dates back to 1967. Syukuro Manabe estimated a 3°C increase in average global temperature if the atmospheric carbon dioxide were doubled. Later estimates—New! Improved! estimates—show roughly the same temperature rise.

The other big input for our engineering committee is geologic history. Lots of natural experiments have already been run, many of them crazier than those we would try on our own. Here is the experimental result: Glaciers advanced all the way to the equator. What was the experiment? My sarcastic line says that God would get a D for His lab notebook. But even without the lab notebook, past climates give us a range of possibilities and a warning about how far the system can change, even without human intervention.

Climate History

The records of ancient climates can be read out of the rocks in two basic ways. The sediments themselves carry part of the record; a glacial "till" deposited by an ice cap is good evidence of a cold climate. Unfortunately, in a burst of enthusiasm during the early 1900s, any rock made of big rock chunks embedded in mud was published as a glacial till. When I was a student in the 1950s, my thesis adviser (Franklyn Van Houten) had a hobby of visiting those supposed glacial deposits. Many of them turned out to be volcanic mudflows, which have a similar texture, but the included boulders would all be volcanic rocks.

Fossils also help, but in an interesting way. Animal fossils (including invertebrates) are great for telling time, but animals are pretty clever at adapting to different climates. Plant fossils don't evolve as fast as animals, but (as all gardeners know) plants are very sensitive to climate. For the last fifty years, paleobotany, the study of ancient plants, has not been particularly fashionable. The "dean" of American paleobotanists is Andrew Knoll at Harvard, but he's a leader with only a small group of followers.

The most spectacular feature in the ancient climate record is not of much use to us because we don't understand it. Not long before the emergence of multicelled animals at the base of the Cambrian there were at least two glacial episodes that apparently froze Earth to the equator. Many of the early climate computer models did the same thing: The entire ocean iced over and the continents were permanently snow covered. Manabe said, "Earth would look like golf ball." The geologic period from about 630 to 850 million years ago has been named the "Cryogenian." The Cryogenian glaciations were probably the biggest climate catastrophe in Earth's history, but until we understand how we got into and out of those glaciations, we cannot base policy on the interpretation.

After the beginning of the Cambrian (542 million years ago), Earth seems to have had two states:

- The "greenhouse state," with no major glaciations and only modest temperature differences between the equator and the poles. There is an exception: If continental drift carries a major continent across the North or South Pole, there can be localized glaciation. North Africa during Ordovician time is an example. The greenhouse comprises about 90 percent of the time since the base of the Cambrian.

- The "icehouse" is the remaining 10 percent. There were only two major post-Cambrian icehouse times: during the Pennsylvanian and Permian, roughly 270 to 330 million years ago, and during the Pleistocene, the most recent 1.8 million years. The Pleistocene glaciation has not ended. Today is the first day of the rest of the Pleistocene.

When Earth is in its icehouse mode, glaciation comes and goes in cycles, which I will discuss later.

An important question for our engineering committee is the cause of greenhouse versus icehouse. There are lots of suggestions, but here's my favorite. Today, the world is divided into two basins: the Atlantic and the Indo-Pacific. As far as water is concerned, high mountain ranges pull most of the moisture out of the air wherever the wind crosses the mountains. There are two water connections between the Atlantic and Indo-Pacific basins: the sea-level connection in the Southern Ocean, and water vapor carried by the trade winds across the Isthmus of Panama, where there are no high mountains. The water vapor crossing Panama is derived from evaporation in the Caribbean Sea—part of the Atlantic Basin—and the water rains out over the eastern Pacific. A salinity contrast builds up between the saltier Atlantic and the less salty Pacific and Indian oceans, and the stirring in the Southern Ocean limits the size of the salinity contrast.

The salinity contrast has a major effect on climate. When chilled, salty Atlantic water becomes the densest water anywhere in the world ocean. The world ocean fills up with dense Antarctic Bottom Water, and the uppermost part of the ocean acts like a shallow

Ocean Surface Salinity

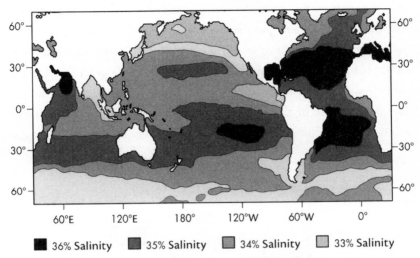

■ 36% Salinity ■ 35% Salinity ▨ 34% Salinity ☐ 33% Salinity

Salinity (dissolved salt) at the ocean's surface results from water vapor transport by evaporation and precipitation. The largest feature is high-surface salinities at 30° north and south resulting from intensive evaporation and little rainfall, which also creates the great desert belts on land. The variation from that simple pattern shows up as high salinities in the north Atlantic and low salinities in the north Pacific, caused by the trade winds carrying water vapor across the mountain-free gap at Panama. Atlantic water, when chilled, fills the oceans below 300 feet with dense seawater and inhibits oceanic heat transport away from the equator. (© Steve Deffeyes)

lake. The less salty Pacific water will not sink even when chilled to its freezing point. The Pacific does not have a Gulf Stream; the Kuroshio Current moves eastward from Japan. In the Atlantic, the Gulf Stream works not because it is warm; it works because it is salty. Chill the salty water and it sinks and pulls more Gulf Stream water up from the Gulf of Mexico.

And when did the two-basin story begin? North and South America had been unconnected until the Isthmus of Panama emerged from the water. The emergence of Panama is usually placed at the "beginning of the Pleistocene," the change from greenhouse to icehouse. It wasn't a sharp change; probably the Panama land bridge emerged gradually and the Pleistocene icehouse took over gradually.

The previous Pennsylvanian-Permian icehouse period existed for a long time. If it is 270 to 330 million years, that's 60 million years compared to 1.8 million (so far) in the Pleistocene. A clever evaluation of the start of the earlier icehouse places it right at the time when the west end of the major Tethys seaway closed. Aha! That's

Late Paleozoic Continents

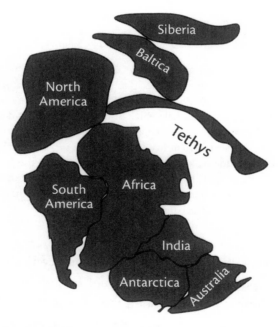

An east-west seaway named Tethys existed from about 300 million years ago to 100 million years ago: long enough to develop a distinctive population of marine fossils. Closure of the western end of Tethys may have caused Tethys to become salty like today's Atlantic, resulting in a cyclic series of glaciations.

an earlier time when Earth was divided into two basins. So today we're in the icehouse mode. If the Pennsylvanian-Permian icehouse is a hint, we may remain in an icehouse environment for millions of years.

Glacial Cycles

During a long icehouse era, glaciers typically come and go in cycles. Glaciers usually grow slowly and retreat rapidly, giving a sawtooth pattern.

Glacial cycles take tens of thousands of years to come and go. The astrophysicist Milutin Milankovitch (1879–1958) was the first to link glacial cycles to cyclic changes in Earth's orbit. The orbital cycles are not mysterious, although the explanation involves some

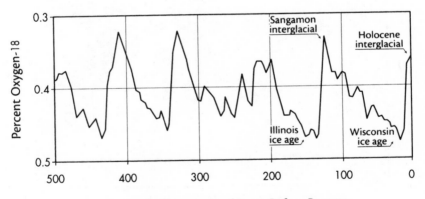

Age in Thousands of Years Before Present

The lighter of the two naturally occurring isotopes of oxygen (O^{16}) evaporates more rapidly from the ocean than the heavier O^{18}. If large amounts of lighter oxygen are stored in continental ice caps, the remaining seawater is enriched in O^{18}. The amount of O^{18} retained as calcium carbonate in oceanic fossils gives us a record of the amount of ice stored in glacial ice caps. Ice age cycles take about 100,000 years each. (© Steve Deffeyes) industrial-strength math. The Milankovitch cycles have periods of 23,000, 41,000, and 100,000 years. Our most recent glaciations, four of them, have a 100,000-year repetition time. Why is the 100,000-year period expressed more strongly than the other two? My candidate for the obvious answer: The world's rivers take 100,000 years to replace all the dissolved oceanic carbon dioxide (which occurs in seawater as bicarbonate, HCO_3^-). Gas bubbles trapped in the Greenland and Antarctic ice caps contain less carbon dioxide than the preindustrial atmosphere. I'm happy with the idea that oceanic bicarbonate swings in tune with the 100,000-year orbital cycle.

The most recent glacial cycle was at its peak 17,000 years ago

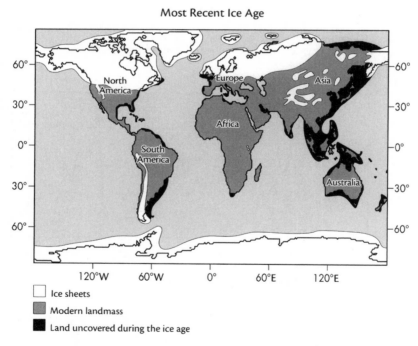

Most Recent Ice Age

☐ Ice sheets

▨ Modern landmass

■ Land uncovered during the ice age

The major ice caps at the height of the most recent glaciation (17,000 years ago) were largely in the Northern Hemisphere. The southernmost edge of the North American ice cap falls halfway between Rutgers and Princeton in central New Jersey. (© Steve Deffeyes)

These overhanging rock shelters at Les Ezyies have probably
been occupied continuously for 30,000 years. Until
10,000 years ago, Les Ezyies was close to the southern edge
of the continental glacier.

and had its most rapid retreat around 10,000 years ago. At the
17,000-year peak, glaciers around the world had advanced, and the
biggest continental glaciers were around the north Atlantic. It came
as a surprise to me that Cro-Magnon cavepersons lived in Europe
only forty miles from the edge of the continental ice sheet. The first
Cro-Magnon skeletons discovered in Europe were found in central
France at Les Ezyies (pronounced "lay say ZEE"). The earliest
skeletons and tools from Les Ezyies are dated at 30,000 years before
present. The Cro-Magnon people had already displaced the earlier

Neanderthals from most of Europe. The Cro-Magnons had skeletons essentially identical to ours. They had stone spear points and fire, hunted big game animals, and probably made fur clothing. Even so, surviving that close to the glacier front had to be difficult. However, it is a strong hint that, as a species, we could survive a major glacial cycle.

The retreat of the continental glaciers occurred from about 12,000 years ago until 5,000 years ago. Today, we are left with the Greenland and Antarctic ice caps as souvenirs from an earlier time. Remember, geologists consider anything shorter than a million years to be just pocket change. Agriculture and animal domestication began about 10,000 years ago. It is interesting that fig trees might have been the first plant to be grown on purpose; you can break a large twig off of a fig tree, poke it in the ground, and it will grow. The onset of agriculture had a huge environmental impact: faster soil erosion and deforestation. From the climate point of view, the years from 900 to 1300 may have been a period of consistent warmth. The *Domesday* book inventory, from the year 1086, listed forty-six places where wine grapes grew in southern England. The Little Ice Age from about 1350 to 1850 made it impossible to harvest wine grapes in England. In his "Season on the Chalk," John McPhee described the renewed production of sparkling white wines in southern England.

The warm medieval period is also reflected in the building of the great Gothic cathedrals in France. Apparently, there were enough agricultural surpluses to allow society to put serious effort into building cathedrals. The first major Gothic cathedral was started in Laon in 1160. Both the building and its site are spectacular. I claim that it must have been like covered sports stadiums are today; each major city tried to build an even taller cathedral. The tallest intact cathedral is Amiens, started in the year 1220. Beauvais, which began construction in 1225, hit the limit. After two collapses, construction ceased, and only the choir section, transept, and seven chapels of Beauvais are intact today. The onset of the Little Ice Age then pushed the societies down into survival mode; the great French cathedral race was over.

Engineering Agenda

It's time to summarize the agenda for our engineering committee:

- What have we got? A coupled ocean-atmosphere-astronomical-agricultural-industrial system and a partial understanding of how it works.
- How does it work? The typical assumption is analogous to the accelerator and steering wheel of a car; a small change in input is followed by a small change in output. But Wally Broecker has warned repeatedly that the headlight switch in the car is a click-click toggle switch; it's either fully on or completely off. An example is switching the Gulf Stream in the Atlantic to a Kuroshio-like Pacific geometry. We might hear a loud *click* in the middle of the night and wake up to find the Gulf Stream flowing from New York to Spain.
- What do we want? I claim that we don't want the Little Ice Age; we certainly don't want the big ice age of 17,000 years ago. (Try writing an environmental impact statement for burying New York, London, and Stockholm under mile-thick sheets of ice.) We'd like a comfortable living for X number of people, with X yet to be determined. We can shop through the geologic record and pick a target that we want. I think the late Miocene looks interesting.

What do we engineers have for methodology? The following list is by no means complete; it just hints at the range of available choices.

- Restrict fossil-fuel burning. Petrochemicals from oil are okay, clean coal would work if the carbon dioxide can be sequestered, biofuels are fine. Go nuclear.
- Some of the sulfur from volcanic explosions reaches the stratosphere and gets converted to sulfuric acid droplets, which reflect sunlight. The global temperature drop is sometimes called volcano weather. The most spectacular example was the 1815 eruption of Tambora in Indonesia. The year was known in the Northern Hemisphere as "1800 and froze to death." How about a fleet of supersonic passenger aircraft plowing through the stratosphere, burning high-sulfur jet fuel?
- Removal of carbon dioxide by photosynthesis in the shallow ocean is limited in some places by a shortage of iron as an essential nutrient. The suggestion is to add iron to those places, although managing it is still in doubt. The technique is known as Geritol, a name borrowed from GlaxoSmithKline. (My Oklahoma relatives consumed lots of chicken livers, red kidney beans, blackstrap molasses, and pork chops; they didn't need iron supplements.)
- Chemical or physical extraction of carbon dioxide directly from the atmosphere.
- Suitably managed, contrails (condensation trails) from aircraft could help reflect incoming solar radiation.
- Nominations are open from the floor. There are probably lots of other opportunities.

The preceding list is focused on short-term remedies for global warming. There is a larger issue about the greenhouse-icehouse dichotomy. If 90 percent of geologic time is in greenhouse mode, why do we need to be stuck with the icehouse and its glacial cycles? If the Atlantic–Pacific salinity difference causes climate extremes, could we fix the problem? Now don't get squeamish. We engineers have strong stomachs. We dig two sea-level canals through Panama. There are large tides on the Pacific side of Panama, almost no tides on the Atlantic side. We install flap gates to make one canal flow from Pacific to Atlantic at high tide and the other to flow from Atlantic to Pacific at low tide. My calculation says that we could wipe out the Atlantic–Pacific surface salinity difference in five hundred years. "A man, a plan, a canal, Panama." Make that two canals.

Engineering our way out of the fossil-fuel greenhouse effect depends on finding places where a small input generates a large output. Otherwise, the huge size of Earth's climate system means we cannot afford to make changes. Adding tiny amounts of iron to selected places in the ocean is an example of a small cause that could have a large effect.

Aircraft contrails are another example. Layers in the atmosphere, usually above 26,000 feet, can be supersaturated with water vapor. Supersaturation can exist because nucleating new ice crystals is a slow process, although natural cirrus clouds show that spontaneous nucleation does happen. If an airplane flies in a supersaturated layer, the water from the engine exhaust condenses into ice crystals. Those tiny ice crystals can bump into one another, generating yet more ice nuclei. The initial narrow contrail will widen with time. In the extreme case, the contrails can fill the entire sky, a condition known as aviation cirrus.

Contrails have two effects on climate. When the sun is high in the sky, the white contrails reflect sunlight out into space, generating a cooling effect. At low sun angles, such as during the winter at high latitudes, or at night, the contrails trap outgoing infrared rays from the ground surface, a warming effect. If we are concerned about

Although contrails from aircraft usually evaporate about a mile behind the airplane, if an atmospheric layer is supersaturated with respect to ice, then the contrail can persist and widen. In this photograph, the freshly formed contrails are narrow and the older contrails have widened into long cloud streaks. (iStockphoto)

global warming, we would like to encourage contrail formation during summertime at low latitudes, and discourage contrails at night and during the high-latitude winter. The atmosphere is usually layered, and moving the airplane up or down can create or avoid contrails. The B-2 stealth bomber avoids tattletale contrails by choosing appropriate altitudes. Compensating passenger and freight airplanes for choosing slightly less than optimal elevations could accomplish a lot at a modest cost.

At the moment, it isn't optimal to have long arguments against any one single climate-control scheme. It helps to get a lot of possibilities on the table for discussion and try to find one that will work at reasonable cost.

NINE

Natural Gas

N atural gas has long been the poor cousin in the petroleum industry. Before the 1970s, natural gas either was a waste product that was burned, or was sold for a much lower price than oil for the same energy content. A new development, called shale gas, might constitute a rebirth of the natural gas industry. It's too early to know for sure, but the new possibility is explained in the last half of this chapter.

We know more about future oil than we know about future natural-gas supplies. For decades, *Oil & Gas Journal* has published annual country-by-country numbers for oil production and reserves, but for natural gas they report only reserves, not production. Even today, a lot of natural gas is flared—burned off as a by-product of oil production. Is that flared gas to be counted as "produced" or not? We have to admit that we simply do not know how much natural gas has been produced.

In an attempt to fill this gap, Matt Simmons compiled a small subsample of the world gas history. He sent some of his bank staff in Houston on an errand to the Texas Railroad Commission in Austin.

The bank collected a history of natural-gas drilling and production for fifty-three Texas counties. It turned out that gas wells are living shorter and shorter lives. It sounds like a medical emergency, but actually it is good news. Improved techniques for hydrofracturing treatments ("frac jobs") have increased the rate of production. It takes less time to deplete the gas reservoir. Investors in the well like to get their money back sooner rather than later. Producing slowly is not a virtue.

Burning natural gas does add carbon dioxide to the atmosphere, but much less than burning oil or coal. Natural gas is versatile. It is widely used to generate electricity, there is an increasing use of natural gas to drive taxicabs and bus fleets, and natural gas is currently the cheapest source of hydrogen for manufacturing nitrogen fertilizer and for upgrading heavy conventional oil. It is possible to convert natural gas into liquid fuels. Could natural gas be the savior of our energy problem?

Here's the puzzle: Will natural gas be available, as an alternative to coal and uranium, until we eventually have to go entirely solar? If we redesign our energy economy to depend on natural gas, will it run out when we are only partway toward our goal? There have been several boosts in the availability of natural gas in the United States. I'll take them in their historical order.

Below the Oil Window

The oil window is the depth range from 7,500 to 15,000 feet. Any rocks that have ever been buried deeper than 15,000 feet will have had their oil cracked down into natural gas. (As mentioned earlier, in a few places salt bodies protect oil even below 17,000 feet.) Until the early 1950s, natural gas sold for less than three cents per thousand cubic feet. Drillers looking for oil learned to avoid areas that had been, or were today, deeper than the oil window. Today, natural gas sells for more than \$3 per thousand cubic feet. Increasing the price by a factor of one hundred is a morale builder. Suddenly, all those natural-gas-only terrains were profitable targets. The "oil boom" of the early 1980s was actually a gas boom.

Tight Sands

Reservoir rocks, for oil or for natural gas, have to have both porosity (open pores in the rock) and permeability (open connections between the pores). Natural gas has a viscosity (a resistance to flow) that is less than the viscosity of crude oil by a factor of one hundred. Low-permeability sandstones that would be uneconomic oil reservoirs can be profitable for natural gas. Frac jobs can enhance further gas delivery from low-permeability "tight" sandstones.

In addition, in several areas these tight sandstones are gas saturated over extensive areas, hundreds of miles across. This is another game changer. Instead of worrying about gas floating above water in a trap, the local search is for higher-quality sand thickness and sand permeability. It's like shooting fish in a barrel. During the 1980s, programs that I advised in New York and Pennsylvania drilled one hundred consecutive gas wells without a dry hole.

Coal Gas

Beginning in the late 1700s, some—but not all—coal mines were found to be dangerous because natural gas leaked out of the coal and caused fatal explosions. Around 1990, a technique emerged for drilling holes into unmined coal beds to extract natural gas. The initial sweet target was in the northern San Juan Basin, along the

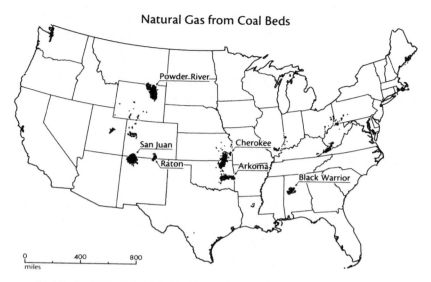

Natural Gas from Coal Beds

Coal beds that produce commercial amounts of natural gas are distributed around the United States, but some important coal-producing areas like Appalachia and Illinois are not major natural gas producers. (© Steve Deffeyes)

Colorado–New Mexico border. The next target to emerge was the Powder River Basin of Wyoming, where there are coal beds 100 feet thick.

Among explorers for oil, buried plant debris has a reputation for producing lots of natural gas and almost no oil. (Really rich source rocks for oil come from marine algae; more about that in a later section.) Coal is derived from plants, mostly coastal swamps buried during a rise in sea level. Natural gas in coal is different in kind from traditional oil and gas reservoirs. Instead of finding oil and gas in pores in the rock, the gas in coal is adsorbed to the organic matter inside the coal. (Historical note: Rosalind Franklin, the scientist who clarified gas adsorption in coal, went on to do X-ray studies of DNA. It was her results that were the initial clue used by Watson and Crick.) The natural porosity in coal largely consists of fractures, called "cleat" by coal miners. In some areas, a hole drilled into a coal bed initially produces water from the fractures. The producer has to have patience, and deep pockets, to pump water out of the fractures (from months to years) before the natural gas starts to emerge from the interior of the coal. Today, about 10 percent of the U.S. natural-gas production comes from coal beds.

Shale Gas

Although shale gas is the Next New Thing, it actually has a longer history than the oil business. Commercial natural-gas recovery from shale dates back to 1825; the celebrated Drake oil well was drilled in 1859. There was a known site where flammable natural gas was bubbling from beneath a small stream near Fredonia, New York. A shallow well was dug by hand—pick and shovel—into the rock. Natural gas was piped through hollowed-out logs and used to operate gaslights. On his last tour of the United States, in 1825, General Lafayette saw the streets of Fredonia lighted at night. I made a similar trip; the gaslights are no longer active, but there is a plaque commemorating the event.

Shale is also known as mudstone, a sedimentary rock so fine grained that the adsorbed water is essentially part of the rock. However, mudstones with 8 or more percent of marine organic matter generate oil when buried into the thermal oil window. If the organic-rich shale is buried below the oil window, natural gas is generated and a portion of the natural gas can remain within the shale. Although nobody knew it in 1825, the natural gas in Fredonia came from a mature oil source rock, the Marcellus Shale.

In the late 1990s, the first whiff of the renovated shale-gas industry was in North Texas. A Texas oilman, George P. Mitchell, recognized that there typically were "shows" of small amounts of natural gas whenever a well drilled through the Barnett Shale.

Natural Gas from Shales

Mature oil source rocks, those that have already delivered their oil, are a new and important source of natural gas. An old-fashioned land rush developed to lease potentially productive areas. However, the wells are so expensive that low natural-gas prices could make much of the area uneconomic to drill. (© Steve Deffeyes)

Mitchell spent several years, and a bucket of money, learning how to extract commercial amounts of natural gas from the Barnett Shale. The eventual winning combination was drilling horizontal wells through the shale and hydrofracturing to open cracks from the well bore into the shale. Hubbert emphasized that the ease of fluid flow (gas, oil, or water) changes with the grain size squared. The rate of fluid flow in a sandstone with millimeter-sized sand grains is a factor of a million faster than flow through a shale with micron-sized grains. Mitchell's successful struggle was to get a commercially valuable flow rate out of a fine-grained rock.

Horizontal Drilling

Horizontal drilling has become an important part of the oil and gas business. I happened to be present at an early (and unsuccessful) test in the 1950s. It was already known that the first few feet surrounding a well bore constituted an important limitation to the flow of oil or gas into a well. John Zublin and his son developed a flexible curved section of drill pipe to drill about 30 feet into the surrounding rock. The test was at the Northwest Lake Creek oil field in Wyoming. Unfortunately, the Zublin gadget broke off as drilling progressed (it's still down there). Instead, the techniques for drilling inclined holes evolved into horizontal holes. A landmark was the mid-1980s replacement of the steel pipeline and water pumps alongside the Bright Angel Trail in the Grand Canyon. A directional well began at the canyon rim and was deviated to emerge in the canyon at the Indian Gardens springs. Today, wells that begin vertically at the surface can be diverted to go completely horizontal in the producing zone. Sensors adjacent to the drill bit relay information so that the driller can guide the bit to follow the oil- or gas-productive horizon.

Another possible improvement in drilling technology is the coil-tubing drilling rig. Instead of drilling with threaded steel pipe, the coil-tubing rig uses an enormous reel to store the pipe. There are reports of coil-tubing rigs drilling 3,000-foot vertical wells in one day: bringing in the rig, drilling the well, and moving the rig out in twenty-four hours. Technology for deviating a vertical surface hole to a horizontal hole several thousand feet long has been developed. Only two companies are producing the materials for coil-tubing drilling; both are in Houston. A third company is coming on line. Guess where? The Houston area. In fact, a rush of new technology is emerging for coil-tubing drilling and well completion. The

United States has taken the lead. There is an online primer about drilling with coil tubing at www.icota.com.

About every fifteen years, someone reinvents the slim-hole drilling strategy. Especially for natural gas, pipe one inch in diameter can carry commercial production. It's happening again with coil-tubing rigs. Previous slim-hole strategies failed because the mud circulation worked better in a seven-inch-diameter hole.

Frac Jobs

The other half of the technology is hydrofracturing. A row of large truck-mounted pumps force water into the well at pressures and rates so fast that new cracks are opened. In the water are sand grains to prop open the newly formed fractures. The technique goes back to the 1950s, when it was realized that the pumper trucks used to pump cement into wells were powerful enough to open cracks. The cementing technique was developed by Erle P. Halliburton, founder of the Halliburton Company. Although there are competing companies with equally powerful equipment, I still think of the Halliburton name as a metaphor for a frac job.

In 1957, the theoretical understanding of hydrofracturing was advanced in a paper written by M. King Hubbert and David Willis. Hubbert and Willis pointed out that a newly formed hydrofracture will open in the easiest direction—the direction that requires the least fluid pressure to crack the rock. In 1957, the theory of plate tectonics was still in the future, but today the stress directions and the easiest fracture orientations have been measured across most of the world. In the United States east of the Rocky Mountains, the easy-opening fractures point toward the Iceland hot spot. The most common interpretation (which not everyone agrees with) says that upwelling material from the lower mantle comes up beneath Iceland, generates abundant lava flows, and spreads out in all directions beneath Earth's brittle outer crust. The compressional stress from the flow away from the Iceland hot spot leaves the least-stressed directions to be pointed toward Iceland. If you don't like the interpretation, then "Iceland" is just a way of remembering which way the artificial fractures will open.

A corporate white paper from Schlumberger pointed out that there are two competing strategies for fracturing horizontal wells:

- Drill the horizontal portion of the well pointed toward Iceland and try to open one long fracture the length of the well.
- Drill the horizontal well 90 degrees away from Iceland and try to open a series of independent fractures perpendicular to the well.

A choice between the two strategies will come from field experience. The crucial piece of the puzzle is that frac jobs on long holes can cost more than drilling the well. Lower-cost fracturing would open up substantial additional areas for exploration.

Generalization

The breakout for shale gas occurred in the years around 2004. The five-word sentence was "Mature source rocks produce gas." It was a quick generalization that opened a whole category of drilling targets. It narrowed the search from all shales—which are the most common sedimentary rocks—to a subset of shales rich in organic matter. As the organic-rich shales are buried to depths greater than about 7,000 feet, the organic matter starts to break into smaller molecules, which is crude oil. Beyond 15,000- to 17,000-foot depths, the temperature is high enough to break the oil, both inside and outside the shale, down into the smallest organic molecule. The smallest molecule is methane, CH_4, which makes up most of the natural gas. The five-word sentence stated that there would usually be recoverable natural gas trapped in the organic-rich shale.

Lots of mature source rocks had been identified long before 2004. (An extended explanation of the origin of source rocks is in chapter 2 of *Hubbert's Peak*.) Identifying source rocks was more than just an academic exercise. Obviously, a province without mature source rocks is not worthy of further exploration. More subtly, oil and gas migration out of the source rock will interact with geologic structures that develop over time. As an example, there has been speculation that some oil in the Powder River Basin (northeast Wyoming) actually migrated from sources in the Big Horn Basin (northwest Wyoming). If the migration happened, it had to occur before the uplift of the mountain range separating the two basins.

Most oil companies quickly compiled lists of known mature source rocks. Some of the source rocks have local names but may be part of a larger target. For instance, there are nine source rock names in the Rocky Mountain region, all from the middle of the Cretaceous (about one hundred million years ago). Likely, the nine names are synonyms for one or two regional units.

A single oil province can have multiple source rocks. A spectacular example is the Paradox Basin in the Four Corners section of Utah. The Paradox Basin was an isolated arm of the sea at the time when repeated glacial episodes in the Southern Hemisphere caused sea level to flap up and down. Whenever sea level went down, seawater evaporated to form salt beds. As sea level came back up, less dense seawater covered the heavy salt brines. No oxygen could mix into the salt water, and all the organic matter that fell to the bottom of the basin was preserved to form an organic-rich source rock. The source rocks are so abundant and so rich in organic matter that half of all the porous rocks in the Paradox Basin are full of oil.

One of the Canadian source rocks is called the "Poker Chip." I had no trouble translating that: A core drilled from an oil well is a cylinder, and the thin sedimentary layers in the organic-rich shale cause the core to break into black disks that look like poker chips.

Once you know that oil source rocks exist, it is not difficult to find them on the ordinary well logs. The primary indication is the resistance of the rock to the flow of electricity. During the early 1920s, downhole electrical resistance measurements were one of the foundations that Pierre and Marcel Schlumberger used to build their company. Within a stack of ordinary shale, the oil source rocks stand out as having higher electrical resistance. A secondary help comes from a downhole radiation counter. Typically, a few parts per million of uranium are found in the organic-rich source rock. The uranium is there because the oxygen-poor environment that preserved the organic matter also reduces water-soluble oxidized uranium into an insoluble form.

As the map on page 102 shows, there are about twenty areas being actively explored for shale gas in the United States. Halliburton estimated the total recoverable shale gas might be one thousand trillion cubic feet, equivalent to a fifty-year supply at our present usage rate. Gas from coal beds and shale gas represent a game-changing opportunity. It won't fully solve our energy problems, but it could be an important help.

The expanded use of natural gas is not a newcomer. *Methane: Fuel for the Future* (1982) summarized a meeting organized by the government of British Columbia. The time–depth chart from my article wound up on the front of the dust jacket. Even in 1982, natural-gas-powered automobiles were operating from New Zealand to Italy. In the United States today, owners of vehicle fleets often operate their own natural-gas fueling stations, but there are only a few stations with public access.

Here we have a chicken-and-egg problem. Although building a natural-gas-powered automobile costs no more than building a gasoline car, the automobile companies won't market natural-gas cars because there are no filling stations. Nobody will build filling stations because there are no natural-gas cars. This is a situation where a governmental boost might get the ball rolling.

Natural gas is readily transported on land through large pipelines, but oceanic gas transport requires huge investments in refrigerated ships and loading/unloading facilities. This divides the natural-gas market into continent-sized units. Today, Eurasia is endowed with huge conventional natural-gas fields in Qatar, in Iran, and in western Siberia. There has been lots of news coverage of Russian attempts to dominate the natural-gas supply for Europe. When oil and gas prices were near their peak in the first half of 2008, Gazprom in Russia signed some contracts in the Caspian area (what I call the "-stans") for high-priced gas. This may be a continuing struggle, so stay tuned.

As the shale-gas drama unfolded, I wondered about the source rocks for the Middle East oil fields and the North Sea. Both areas have thick, rich source rocks, both of them in the Upper Jurassic. However, drilling for shale gas is considerably more expensive than conventional wells in the porous gas reservoirs. For the moment, there is little or no economic motivation to look for shale gas in the Middle East.

The shale-gas drama came down to a personal level. My brother and I each inherited pieces of Oklahoma farmland from our mother. A company leased our oil and gas rights and was preparing in 2007 to drill a 17,000-foot gas well. They even built a road into their proposed drill site and built a multiacre gravel drilling pad on my brother's property. Then, abruptly, they abandoned the well and announced that they would not renew the lease when it expired. Probably, they went off chasing shale gas. Our farms are adjacent to U.S. Highway 81, which is the route of the historic Chisholm Trail. I tried to convince my brother to open the gravel drill pad as the Chisholm Trailer Park. He didn't think it was the least bit funny.

I mentioned that the federal government might help solve the chicken-and-egg problem about natural-gas-powered automobiles. There is an additional possible government move. Matt Simmons proposed establishing a minimum support price for oil and for natural gas. The threat of lower prices—such as happened in the 1990s—makes companies hesitant to initiate price-sensitive projects. The government could pick reasonably low support prices for oil and natural gas and announce that it would buy oil or gas on the open market if the price went below the support price. The government would be free to sell the oil or gas when the price was higher. Buy low, sell high. It would remove an enormous uncertainty from investment decisions.

We have to remember that natural gas, like oil, is a finite resource. It runs out someday, but it helps us while we adjust for a renewable-energy future.

TEN

Oil from Algae

I better confess this first: I'm attracted to the oil-from-algae schemes because most of our natural crude oil comes from the remains of single-celled marine algae. The U.S. Department of Energy had a substantial program evaluating pond-raised algae as a biofuel source. When crude-oil prices dropped in the 1980s, the DOE put the kibosh on the algae program. Today, entrepreneurs are reading the old DOE reports as a guide to starting up cultured-algae programs.

Most of today's existing biofuel programs divert food crops: sugarcane, corn, oilseeds, palm oil, and soybeans. This is only one of several reasons why food prices have risen as natural crude oil became scarce. Tide flats, bogs, and shallow lakes potentially could be converted to algae farms. I know, I know, every time we nominate a specific site either the bird lovers will identify it as the breeding ground of the pink-cheeked pushover or my fellow First Americans will claim it as a sacred tribal burial ground. However, we better not locate the facility on the Brooklyn waterfront; these ponds may smell like an elephant's hind foot. There are both freshwater and marine algae to choose from.

Three major groups of marine organisms could generate oil: cyanobacteria, diatoms, and dinoflagellates. The cyanobacteria, which used to be called blue-green algae, are a seriously ancient group with a fossil record extending back for billions of years. Diatoms build exquisitely beautiful microscopic skeletons of silicon dioxide; commercial diatom ponds would require some form of silicon as a fertilizer. Dinoflagellates usually get into the news because they cause toxic red tide blooms in nutrient-rich seawater. I'd as soon not deal with dinoflagellates. The most attractive feature of oil-from-algae is fuel production per acre, at least a factor of three larger than the standard biofuel crops.

Biological oils, fats, and waxes—lipids—all share a similar chemistry. Lipids have a water-loving head (called glycol) attached to water-hating hydrocarbon chains. A double layer of lipids, with the hydrocarbons in the middle and glycerine groups on either side, makes an effective divider to be used as an outer cell wall or as an interior divider inside the cell. We want to separate the hydrocarbons; the glycerine is left as a by-product.

On the Internet, there are kitchen-scale recipes for reacting natural lipids with small amounts of methyl or ethyl alcohol to produce biodiesel. The recipes sound to me like a quick way to burn down the house. I'm not trying it at home. On an industrial scale, the same reaction produces biodiesel. A substantial portion of the land area of Malaysia has been converted to oil palm plantations. One of the Malaysian palm-oil-producing companies, Kuala Lumpur Kepong Berhad, owns the Crabtree & Evelyn cosmetic firm to sell part of the glycerine by-product. Glycerine (also known as glycerol) occurs naturally in a number of foods, most spectacularly in the expensive late-harvest Sauternes dessert wines. However, the increased demand for palm oil as the least expensive cooking oil has some of the newly built biodiesel refineries standing empty. My hope is that algal cultivation would not compete with food products.

The inputs to an algal cultivation pond are sunshine, carbon dioxide, and phosphate and nitrate fertilizer (plus some form of silicon if diatoms are the product). It is possible that the carbon dioxide, phosphate, and nitrate could come from urban, agricultural, or industrial sewage. Another possible source is the carbon dioxide from fossil fuel burning.

There are two exciting new possible ways to utilize cyanobacteria (see *Nanoscale: Visualizing an Invisible World*, pp. 43, 131). Several plans for growing and harvesting single-celled cyanobacteria were developed from 1975 to 1983 and interest picked up again around 2005. However, two new methods are being developed for using cyanobacteria without collecting and destroying the cells.

- A start-up company, Algenol, with backing from Dow Chemical Company, has genetically modified a strain of cyanobacteria to produce sugar and, in the same cell, to ferment the sugar into ethanol. The modified cyanobacteria can live in water-filled trays covered with clear plastic and the ethanol can be recovered from the vapor over the trays. Actually, this project is not motivated by a need to produce ethanol as a motor fuel; Dow needs the ethanol to produce plastics.
- Craig Venter's company Synthetic Genomics is receiving financial and engineering support from ExxonMobil to modify cyanobacteria to release cell-wall lipids, which can be converted to hydrocarbon fuels. The early news reports say only that Venter has "an idea for getting the cyanobacteria to release lipids." It isn't known yet whether he is getting the cells to overproduce lipids and to spit out the excess, or whether he has some other idea. An obvious guess is that he could skim the excess lipids off the water surface.

Although these two innovations are well funded, rather than sitting back and waiting we could encourage additional competitors to enter the single-cell-algae race. There are hundreds of cyanobacterial species to work on, several chemical products, and lots of biological genes to splice into the cells.

A level reached in the 1980s was a barrel of biodiesel per acre per day. Conventional oil wells, spaced forty acres apart, producing forty barrels per day each (one barrel per acre-day) are considered to be economically successful.

ELEVEN

Transportation Efficiency

On June 6, 2008, readers of the *Philadelphia Inquirer* and the *Philadelphia Daily News* found newspaper ads for a new airline, Derrie-Air, whose fares depended on the weight of the passenger, plus luggage. The ad was a hoax, but like any good hoax it contained internal clues. Earlier I had a similar idea and worked out the theory, but I didn't package it as cleverly. (The website for their fake airline is at flyderrie-air.com.) If the airline had been real, the seats would fill up with skinny computer nerds carrying minimal baggage.

What we need is an overall comparison of energy consumption per mile of payload carried. Of course, "payload" is the transportation that the carrier gets paid for. On an airplane, the fuel is not payload. Fuel is not a profit center for the airline; it's an expense. I suggest using the miles that the payload could be carried by consuming a weight of fuel equal to the payload. For trucks, boats, and airplanes, it is a simple concept: ton-miles of payload per ton of fuel burned. Working the problem in pounds instead of tons gives the identical answer: pound-miles of payload per pound of fuel. However, other transport systems can be brought under the same umbrella. For an electric transmission line, it is the kilowatt-hour-miles moved divided by the kilowatt hours of energy used. For a natural-gas pipeline, it is the cubic feet of gas moved divided by the natural gas burned to run the compressors. The useful feature is that all the different transportation efficiencies are measured in miles. (Metric system purists can multiply all of the mile efficiencies by 1.6 and get kilometers.)

There are some transportation systems that are not easy to translate into miles (or kilometers). For instance, the amazing French TGV trains are electrically powered, and 75 percent of French electricity comes from nuclear-power plants. The TGV is a nuclear-powered train, not easy to convert to payload ton-miles per ton of fuel.

My initial list follows. The numbers are scrounged from all manner of sources. Further, there are suspicions that some equipment manufacturers have shaded the numbers to make their products look good. But we have to start somewhere, so here goes:

Transport System	Miles (payload ton miles/ton fuel)
Concorde supersonic airliner	467
Learjet business jet	3,400
Electric power line	15,600
18-wheel tractor-trailer	18,500
Natural-gas pipeline	25,000
Freight train	63,000
Canal or river barge	160,000
Oil tanker, at 16 knots	1,000,000
Bulk carrier, at 13 knots	1,300,000

We all suspected that the Concorde was inefficient. To me, the real surprise was the efficiency of water transportation. Today, the lowest-cost way to ship goods from the U.S. East Coast to the Upper Midwest is by the Erie Canal, completed in 1823 and enlarged in 1905. In Shanghai, I timed one coal barge each minute going up the Yangtze River. I'm waiting for UPS to wake up and integrate oceanic container ships with container-on-barge traffic (called COB), and use regional trains and then trucks for local delivery.

When coal is mined to generate electric power, there is a choice between generating the electricity near the mine and sending it by electric power lines to the place it will be used, or sending the coal by trains and generating the electric power close to the power demand. The table opposite shows that train transport is more efficient than "coal-by-wire." A close-up view of several of these transportation modes is in John McPhee's *Uncommon Carriers*.

Students: Need an original term paper? Head out to your local highway truck stop. Tell drivers that you are gathering data for a term paper. Explain that you do not want their name or license plate number. Ask them how many miles per gallon they are getting. Then ask about their payload, not including fuel weight or the empty truck weight. They will probably know their payload in pounds (because the highway department truck scales are in pounds). You can convert the diesel fuel weight to pounds using 7.2 pounds per gallon. Include in your term paper both a long table of the data and a graph showing payload weight versus fuel consumption. The data table and graph fill up space, so you need to write fewer words. If your parents are nervous about your hanging out at a truck stop, borrow a lineman from the football team to follow you around and look interested.

A potential flaw in dealing with the global oil shortage is fixing a part of one problem and then congratulating ourselves on having "done something." An example is CAFE standards. I thought they were named cafe standards because they were invented during power breakfasts in Washington restaurants; it's an acronym for corporate average fuel economy. It seems so satisfying to mandate better fuel economy. However, the 1975 version was doubly flawed. Arbitrarily increasing the fuel efficiency pushes increased costs on down to the material suppliers and to the petroleum refiners. Further, the law set a lower standard for light trucks as compared to passenger automobiles. The family station wagon vanished because the SUV could qualify as a light truck. Increasing fuel economy is not simply a freebie; the long chain of consequences has to be considered.

Of the entire world oil production, almost 20 percent is used for transportation inside the United States. Obviously, improving the U.S. transportation system is important. As a start, we need a more detailed "per-mile" table instead of the one shown on page 116. The per-mile contrasts in the table are very large; it isn't likely that someone will quickly design a rubber-tired truck that is as efficient as a freight train.

TWELVE

Metal Deposits

O il isn't the only game in town. Earth supplies lots of other resources: metals, fertilizers, construction materials, agricultural land, and water. All of them are important—and interesting. However, some metals have a long history; the Drake well began producing oil only 150 years ago. It is necessary to see oil in contrast to other mineral resources. I should confess that histories, particularly for metals, are wildly different from one another. Instead of emerging with a single model, we are going to appreciate the diversity of mineral resources.

Gold jewelry dates back to the very beginning of civilization. Grains of metallic gold can be washed from streams and rivers. Gold can be melted at readily achievable temperatures and can be hammered into controlled shapes. It does not tarnish with time; ancient gold jewelry retains its luster even today.

Gold is placed among the noble metals because of its disinterest in participating in chemical reactions. However, "nobility" is a gradual scale, with gold (and the platinum group metals) at one end of the list. Next on the list are silver (it tarnishes, especially with sulfur compounds), and then copper and lead. The noble metals occur in nature either as relatively pure metals or as weakly bonded minerals that are readily broken down to release pure metals. But as we go on down the list, the bonds between the metals and oxygen get stronger and stronger. Iron is the dividing line. Ancient methods, using wood or charcoal fires and primitive air pumps, could produce only an impure iron. Aluminum and titanium come only from modern technologies.

The gradation from noble to nonnoble metals can be expressed exactly by a measure called the electromotive series, in use since the 1890s. (Specifically, the series is the electrical voltage of a half cell, with the metal immersed in a 1-molar solution of dissolved metal. Hydrogen is chosen as the zero point of the series.) The graph below uses the electromotive voltages as the vertical axis and the annual world production of each metal as the horizontal axis.

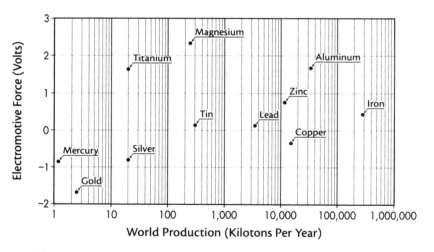

Metals with small, or negative, numbers on the electromotive force scale are readily separated from their ores. Iron and aluminum required serious technology to bring them into large-scale use. Magnesium and titanium, although abundant in Earth's crust, have yet to achieve major usage in industry. (© Steve Deffeyes)

The bottom six metals have been used for four or five thousand years: gold, mercury, silver, copper, lead, and tin. Almost every ore deposit of these six metals, exposed at the surface in Europe, had been known to the Romans. Extensive use of iron began around 1650 and greatly expanded during the Industrial Revolution. Aluminum and magnesium are from the 1900s, and I would rate titanium as one that has yet to achieve its full potential.

Although all of the ancient metals have fascinating histories, it may be most useful to examine tin. The comparative rarity of ore deposits of tin was a sigificant constraint on developing technology. Copper ore deposits occur more frequently. Copper by itself is soft and not useful for making tools. However, adding 10 to 20 percent of tin to copper produces the alloy known as bronze. Bronze can be made into effective tools and weapons. Many ancient cultures go from a Stone Age to a Bronze Age to an Iron Age. Can't have a Bronze Age without tin. Although the official Bronze Age ended about 1000 BCE, it did not end abruptly. About half of the cannons in the American Civil War were bronze; the other half were iron.

Copper has an average abundance of 63 parts per million in Earth's crust; tin makes up only two parts per million. Although a few natural processes enrich average rocks up to ore-grade deposits, the lower abundance of tin assures fewer ore deposits. The natural environment most effective for concentrating tin (and also molybdenum and tungsten) is the high-temperature brines that come off as granite magma solidifies. The brines, escaping through cracks in the rock, deposit crystals of tin oxide (and other minerals) on the walls of the cracks; the deposits are known as veins. If those original tin veins are eroded by weathering, the tin oxide grains can be further concentrated in river deposits, known as placers. The tin mines of Cornwall, England, were in veins; the extensive tin deposits in Malaysia and Indonesia are placers.

Malaysia and Indonesia come closest to each other at the Straits of Malacca, only a mile and a half wide at Singapore, its narrowest point. For a hundred years "straits tin" has been a market name for a particularly pure grade, 99.85 percent tin. In downtown Singapore, I came across a display extolling the history of straits tin, including a tin ingot that I could touch.

The long history of tin use is not over. Today, the two largest consumers of tin are as solder and in various tin-containing chemicals. Solder, for joining electronic parts and copper pipes, was originally two-thirds tin and one-third lead. However, the toxicity of lead is shifting us to soldering alloys that are more than 90 percent tin. The third largest use is in tin cans. The familiar cans in the grocery store are made of steel (iron) sheets plated on the inside with a thin layer of tin. The tin is the rustproof and nontoxic layer that protects the food. It is important to recycle tin cans. The tin is recovered chemically and the underlying iron goes back into steelmaking.

Tin isn't cheap; it sells for about $5 per pound, three times the price of copper. The three largest producers are China, Indonesia, and Peru. I find it fascinating that the widely separated tin mines, with the help of high tin prices, have kept the world moderately supplied for the uses where tin is essential.

In a sense, we have two models. Oil, which gets seriously depleted after 150 years; and tin, which stays in limited supply for thousands of years. In one sense, the price of tin rations it to only the most essential uses. How can we identify the bottlenecks, the choke points, that influence the development of our civilization?

If we go back to the larger picture, the normal expectation would be that we first mine out the richest high-grade deposits. Then we gradually proceed to lower-grade (and larger) deposits. For copper, this seems to be the case. Ancient copper mines contained several percent copper in the ore. Today, most of our copper mines produce ores around 0.5 percent copper. There are rumors of plenty of additional copper deposits that were not quite rich enough to justify mining.

In contrast, lead and zinc mines today are producing from ores as rich as those mined two hundred years ago. The reason: Advances in understanding the origin of lead-zinc deposits have led to successful exploration.

One thread of the story began with the recognition of small but significant amounts of lead and zinc dissolved in the hot salt water produced from oil wells in the Louisiana–Texas area. Hmmm . . . What happens to these subsurface salt brines in the absence of oil wells? They gradually escape as the surrounding sedimentary rock gets compacted during deep burial. How do they escape? The brines can readily flow horizontally; some of the surrounding sedimentary layers are porous and permeable. However, the upward path is hindered wherever there are layers of dense and impermeable rocks, particularly limestones. Where can the metal-rich brines escape across the impermeable layers? Wherever there are caves or collapsed former caves. Lots of metal-rich salt water escaped through those former caves, including the celebrated Tri-State lead-zinc mines of Missouri, Oklahoma, and Kansas. You don't have to hire a mining geologist from the Colorado School of Mines; find a spelunker, a cave freak. Similar lead-zinc mines are being discovered in other areas.

A second thread is an older one about lead, zinc, copper, and silver mines found in deepwater environments. I first heard about these deposits as a student on a 1957 field trip that included the Ducktown, Tennessee, copper mine. One of the professors, Jo Kalliokoski, said to me, "Sederholm claimed that these ores were deposited on the open sea floor." (He was referring to the Finnish geologist Jakob Sederholm.) In 1957, there weren't even hints of the plate tectonic paradigm; Sederholm died in 1934. When the theory of plate tectonics became accepted in the 1960s, one of the key pieces of evidence involved the copper mines on Cyprus. (Copper was named for Cyprus.) A complete section through a slice of oceanic crust and a piece of the upper mantle is exposed on the western half of Cyprus. The uppermost part of the oceanic crust is overlain by deep-sea sediments, and the Cyprus copper deposits are right at the contact between the uppermost lava flows and the over-lying sedimentary rocks.

The final piece of the puzzle fell into place with the discovery of black smoker hot springs near the mid-ocean ridges. I'm the guy who didn't discover the oceanic hot springs. Likely, you didn't either. However, you didn't discover them by accident; I didn't discover them on purpose. In "The Axial Valley" (1970), analyzing the mid-ocean ridges, I explained on page 220 that the hot springs had to exist but they would be very difficult to find. The springs would be local and there are cubic miles of seawater to dilute the hot-spring water. Several searches (not by me) wound up empty-handed. Then, in 1977, blips of seawater only a few thousandths of a degree warmer than the surrounding water were traced to hot springs on the sea floor. Further exploration turned up really hot springs (300°C, the black smokers) that were depositing copper ore.

Normally, we tell science tales by acting as if we were smarter than we really are. Our heavily edited story would begin with the black smokers, then go to plate tectonics, and move on to Cyprus and the many other sea-floor mineral deposits. Don't mention Sederholm; he was way ahead of his time.

In summary, we have a huge variety of metal ore deposits, most of them well described. What we are missing is an inventory for the next round, after our currently active mines are depleted. Here's my "wastebasket" strategy: Mining companies and oil companies work on prospects, places where extraction might be profitable. Once a year, management holds a meeting and various prospects are brought up for discussion. Some prospects make the cut and are scheduled for further investment. The losers wind up in the wastebasket. I desperately want the contents of that wastebasket. It is the possible ore deposits and the possible oil fields that didn't quite make it. If, for example, over a period of years, no company had a vanadium deposit in the wastebasket, we would be warned that a vanadium shortage could occur.

THIRTEEN

Recommendations

Our civilization faces a wide range of serious problems: climate change, health-care costs, a widening gap between rich and poor, water availability, AIDS, freezing of the financial system, budget limitations, and agricultural productivity. Almost everyone feels that his or her most pressing concern should be dealt with first. Unfortunately, we lack the money, materials, skilled workers, and political will to fix everything immediately. This is not prioritizing; it's simply a matter of scheduling. There are some problems with short time scales that must be dealt with immediately; collapse of the banking system is a prime example. Complete conversion to solar energy has a time scale of at least one hundred years. That doesn't mean that we ignore solar energy for the next ninety-nine years: It's an important goal, but we have time to include some research and development.

Climate change is a drama that is played out on a geologic time scale. Right now, the climate-change mandate has not chosen the target we are aiming for. I sometimes accuse the climate-change folks of trying to take us back to the year before the Industrial Revolution. That was 1749, during the Little Ice Age, which ran from 1300 to 1850. Taking us back to the climate of 1250 might be a lot more pleasant. If you want to know more, the key book is Brian Fagan's *The Little Ice Age*. However, I confess that although I enjoyed the pre-1300 part, when the Little Ice Age began, the scene in Europe was so horrible that I stopped reading.

Going back to the time before the Industrial Revolution would not remove all of the major human-caused disruptions. The development of agriculture, beginning about seven thousand years ago, caused a large increase in the erosion rate and modified the geochemistry of several major rivers. We're not about to shut down agriculture to reestablish a "natural" Earth.

Conventional Oil Fields

As a geologist, I am fixated on the crude-oil supply and its effects on the global economy. The world crude-oil market amounts to $1.7 trillion per year. Particularly in transportation and agriculture, we have an enormous investment in a system that runs on crude oil. I'm not arguing that it should be that way; it results from one hundred years of relatively cheap oil. Revising a massive system requires more than passing new laws mandating energy efficiency. In my opinion, world crude-oil production is past its peak; we cannot rescue our transport and agriculture systems by raising world oil production. However, every little bit helps, and here are some possibilities:

- Existing oil fields need to be reexamined for oil-bearing horizons that were not produced under the technology and economics of 1920 to 1970. Contracts are being arranged between Iraq and international oil companies to redevelop specific oil fields, with the company payout to be based on the size of the production increase.

- New provinces, like the deep presalt oil fields in offshore Brazil, are enticing. Most of them require specialized and expensive large drilling rigs. My advice is to encourage these projects, but don't count on them to bring global oil production back up to the 2005–2008 level.

- Minimum price supports for oil and for natural gas would encourage investments in the expensive new fields. I know that this sounds like giving money to the Rockefellers, but the support price level could be set low. If oil began selling for less than $30 per barrel, the U.S. government would buy additional oil for the Strategic Petroleum Reserve. Price supports can have value even if oil prices never drop below the support level. Decision makers planning new projects would not have the nightmare of receiving $20 per barrel for their billion-dollar investment. Oil sands projects are particularly allergic to low oil prices. Matt Simmons has been arguing for minimum prices for several years.

Patching up the existing crude-oil supply doesn't solve the problem, but it relieves a small part of the pain.

Some reports say that China is sitting on $2.1 trillion worth of foreign exchange. They are cutting deals, often structured as loans, with national oil companies for future oil supplies.

Alternative Natural Fuels

Natural gas, coal, and uranium are the remaining energy sources that can be extracted and transported. Each has its problems, but at least we have existing technology for using them.

▪ The recent discovery that mature oil source rocks can produce natural gas is an unexpected and pleasant surprise. The deposits are usually called shale gas (see page 101). However, the immediate response to the new discoveries was a sharp drop in natural-gas prices. Of course, that's the way economics is supposed to work, but I'm worried that low gas prices will strangle my new-born baby. As with oil, we need to post a minimum support price.

▪ Coal is available, possibly on a one-hundred-year time scale (see page 65). Simply burning the coal and exhausting the carbon dioxide into the atmosphere is not likely to be acceptable. My recommendation is to go ahead with coal gasification projects. The technology exists and we need to learn further by doing it: on-the-job training.

▪ Uranium is available, again on a time scale of one hundred years or more. I agree, it's scary. Chernobyl didn't help much. We'll just have to bite the nuclear bullet. Nuclear plants don't add carbon dioxide to the atmosphere.

These three fuels are temporary stopgaps; we don't have to go 100 percent solar next year. However, it is important to recognize that they are temporary. We don't want to get caught in a dead-end street. For instance, even if we were to convince ourselves today that shale gas would solve all our problems, in twenty or fifty years we would again be faced with a difficult problem.

It is outside my range of expertise to list and to evaluate all of the biofuel possibilities. However, one warning: Producing some of the biofuels (such as alcohol from corn) requires almost as much energy as the fuel produces. The immense political pressure to support agriculture can promote marginal schemes.

Mineral Resources

In addition to the energy resources (oil, natural gas, coal, and uranium), there are a number of metals and nonmetals whose prices are going up. Copper for electrical transmission, nickel and chromium for stainless steel, and phosphate for fertilizers are examples.

A new candidate on the list is neodymium, a rare-earth element that is increasingly used in strong permanent magnets. Generators for producing electric power from the wind typically use neodymium-iron-boron magnets. China dominates the world production of rare-earth elements, but rare-earth consumers in other countries are trying to develop alternative supplies.

The use of lithium in lithium-ion batteries is increasing. Historically, lithium was produced from lithium-bearing minerals in pegmatites, coarse-grained rocks. A newer source, salt brines beneath dry lakes in desert basins, has come to dominate the lithium supply.

Of course, agricultural land and water are the most extensive and most important of the natural resources.

Market Volatility

Times of great crisis can be times of important change. This is an opportunity to discard some accumulated baggage in the stock market. I'm not an economist; I'm writing this as an engineer. Like most engineers, I may be reinventing the wheel, but here it is.

An orderly trading market does not need or want wild swings in price. Stock market volatility helps no one except day traders. A recurrent engineering problem is finding the fastest way to get an oscillating system to stop at its equilibrium point. It's called critical damping and it involves introducing friction that is proportional to the speed of the system. My proposal is to introduce a "volatility tax" on stock trades. The tax percentage increases with the rate that the price is changing. If the price remains unchanged for a while, the volatility tax drops to a small minimum value. The tax depends only on the individual stock price history, not on the overall market. And it is symmetrical; abrupt price rises are dampened to the same degree as price drops.

To enjoy critical damping, I rigged a homemade demo consisting of a strong magnet hanging on a string off the edge of a table with an aluminum plate on the floor. If I shortened the string and swung the magnet, the magnet would swing back and forth and eventually stop because of the electrical currents induced in the aluminum. If I lengthened the string until the magnet almost touched the aluminum plate, the magnet would crawl slowly toward its equilibrium point. In between was the length for critical damping. If I let the magnet swing toward the plate, it would sail in and park quickly, almost like magic. Lots of scientific measuring systems are rigged with critical damping to give a stable reading as quickly as possible. A similar behavior might benefit the stock market.

My magnet pendulum had its own preferred time to swing back and forth. The stock market is hit by many forces at irregular times. For instance, ExxonMobil stock is affected by global events, not all of them directly connected to Exxon. Turmoil in Iran is likely to affect the world oil supply, indirectly increase the global oil price, and then raise the ExxonMobil stock price. At any rate, the computer program that sets the volatility tax should be publicly available. The volatility-damping tax would go to a minimum level when the price has been almost constant for a specified length of time. When millisecond trading became public knowledge, I decided that we need a minimum tax to discourage large-volume, small-margin trading.

Of course, when this story emerges there will be quants from

New York to Hong Kong who will look for ways to game the system. Before we abandon the existing system, we better have some tiger teams try to beat the system.

The proposed system equally damps upswings and downswings in price. We abandon all attempts to create a one-way ratchet to cause the stock price always to increase. Also, we do not try to restrain speculators. If I like ExxonMobil stock, I am not asked to limit my stock ownership to match my consumption of gasoline.

A second major change: There is no bias against short sales, which is simply ownership of a negative number of shares. Negative numbers, and short positions, have a long history of seeming spooky. Although negative numbers were used correctly two thousand years ago in China and negative numbers were adopted by Arabic mathematicians one thousand years ago, as recently as two hundred years ago in Europe negative numbers were still a no-no. Even today, many people think of short positions as a sin. A better way to think about it: Short positions have the same function in the stock market that buzzards have in the desert.

Unfortunately, the short-versus-long treatment is not perfectly symmetrical. If you buy and hold stock, you are "long." If the value of the stock goes all the way down to zero, you lose all your original investment. On the upside, your potential profits from stock ownership are unlimited.

In my proposed system for short sales, you put in cash equal to the present market value of the stock. When you close out your short position, you get your original bet back plus the price drop. If the stock value drops to zero, you double your money. If the stock price goes up, the price rise is deducted from your original bet. Here's the asymmetry: If the stock price ever doubles, you lose all your original bet and your short position is automatically closed out.

While we are at it, we could make public what is called the "specialist's book," which is a list of current offers to buy or sell different amounts of a particular stock. Under current practice, the specialist's book is highly confidential. The specialist in a stock is there to maintain an "orderly market," and I'm supposed to be better off if I do not know what buy and sell offers are on the table. Let's put the specialist's book on the Internet in real time and for free.

Some exchange-traded entities, like gold, allow bets on commodity prices without involving the futures markets. I'm not completely convinced that we really need futures markets. Producers and consumers could hedge future prices with simple letter contracts.

In my proposed system, there is no leverage. Sales are for cash only. If your bank, or your aunt, wants to loan you money to play the stock market, that's your business and not mine. Notice that there is full liquidity. Even if the volatility tax is large at the moment, you can still sell stock immediately by offering to sell at the price of the lowest "buy" offer, minus the tax. Who gets the revenues from the volatility tax? The federal treasury gets 99 percent of the tax. And the other 1 percent? That's my royalty fee for inventing the system.

Sustainable World

So what does the world look like a hundred years from now? I'll give two extreme views, but the reader is encouraged to invent other scenarios. Considering various targets, seeing different models, helps us to agree on an option and to work toward it.

- A posting about two years ago in the Peak Oil newsletter suggested that the world would break up into a series of independent small villages, each with green surroundings. Most of the village land would be forest, managed as coppice. I had to look up the word "coppice." It's a strategy for harvesting wood by cutting off trees at ground level, allowing new sprouts to shoot up from the existing roots. A smaller amount of land would be used for gardens, with a few chicken coops thrown in. The villages would not be intellectually isolated. Moving bits is much cheaper than moving atoms. Most of the education would be done over the Internet; you could take an advanced degree in organic chemistry without leaving the village. My guess is that the only people allowed to travel extensively would be exploration geologists.

- Richard Smalley (Nobel Prize for chemistry) had an image closer to our present society. He would gather solar energy (Arizona and Nevada) and wind energy (Wyoming) and transport energy to population centers over lightweight direct-current buckytube transmission lines. In October 2009, the Honda Research Institute, working with Purdue University and the University of Louisville, announced a synthesis procedure that selectively produced electrically conductive buckytubes. Each individual customer would own an energy storage facility—a refrigerator-sized unit down in the basement. Battery-powered plug-in automobiles would provide transportation.

Regardless of the model, a sustainable Earth is going to support only a finite number of people. One of the major disappointments of my life happened in 1967. Linus Pauling (Nobel Prizes for chemistry and for peace) came back to Oregon State University, where he had been an undergraduate. His big public lecture was to be about the carrying capacity for Earth's population. I was poised, notebook in hand, and Pauling announced that he was so upset about the Vietnam War that he could talk about nothing else. I have made several inquiries, hoping that notes for his undelivered population lecture survived. Nothing turned up. My suspicion is that our present-day human population may already be larger than Earth's steady-state carrying capacity. The biggest challenge to our intelligence is managing the transition from a growing economy and a growing population to a sustainable steady-state condition.

If the world were flat, we would have unlimited resources and unlimited space for growth. The world is round.

SOURCES

INDEX

Sources

PREFACE

"Brazil Claims Stake in Oil Find." *Wall Street Journal*, September 1, 2009, http://online.wsj.com/article/SB125174253096373259.html.

Deffeyes, K. S. *Hubbert's Peak: The Impending World Oil Shortage*. Princeton, NJ: Princeton University Press, 2001. (My 2005 prediction is on p. 158.)

Steiner, Benjamin J. "Peak Oil and Policy Recommendations for Reducing Its Effect on the United States Transportation Sector." Senior thesis no. 21236. Mudd Manuscript Library, Princeton University, 2007.

U.S. Energy Information Agency. *International Petroleum Monthly*, August 12, 2009, www.eia.doe.gov/emeu/ipsr/t11d.xls.

ONE   Bell-Shaped Curves

Hubbert, M. King. *Nuclear Energy and the Fossil Fuels*. Houston: Shell Development Company Publication 95, 1956.

MacArthur, Robert, and E. O. Wilson. *The Theory of Island Biogeography*. Princeton, NJ: Princeton University Press, 1967.

Verhulst, P. F. "Notice sur la loi que la population suit dans son acroissement." *Corr. Math. et Phys.* 10 (1838): 113.

Weisstein, E. W. *CRC Concise Encyclopedia of Mathematics*, 2nd ed. Boca Raton, FL: Chapman and Hall, 2003. (The differential form for the Gaussian distribution is equation 65 on p. 1159.)

TWO **Nehring's Critique**

Allaud, L. A., and M. H. Martin. *Schlumberger: The History of a Technique*. New York: John Wiley, 1977.

Archie, Gus E. "Electrical Resistivity as an Aid in Core-Analysis Interpretation." *American Association of Petroleum Geologists Bulletin* 31 (1947): 350–66.

Campbell, Colin J. *The Coming Oil Crisis*. Brentwood, Essex, UK: Multi-Science Publishing Company, 1997.

Deffeyes, Kenneth S. *Beyond Oil: The View from Hubbert's Peak*. New York: Hill and Wang, 2005. (The easy derivation of the logistic equation is on p. 38.)

Hubbert, M. King. "Techniques of Prediction as Applied to the Production of Oil and Gas." In S. I. Gass, ed., *Oil and Gas Supply Modeling*. Gaithersburg, MD: National Bureau of Standards Special Publication 631, 1982, pp. 16–141.

Nehring, Richard. "Giant Oil Fields and World Resources." Rand Corporation Report R-2284-CIA, 1980.

———. "Hubbert's Unreliability—1: Two Basins Show Hubbert's Method Underestimates Future Oil Production." *Oil & Gas Journal* 104, no. 13 (April 3, 2005).

———. "Hubbert's Unreliability—2: How Hubbert's Method Fails to Predict Oil Production in the Permian Basin." *Oil & Gas Journal* 104, no. 15 (April 17, 2005).

———. "Hubbert's Unreliability—3: Post-Hubbert Challenge Is to Predict Production." *Oil & Gas Journal* 104, no. 16 (April 24, 2005).

Simmons, M. R. *Twilight in the Desert*. Hoboken, NJ: John Wiley, 2005.

Sneider, R. M., and J. S. Sneider. *New Oil in Old Places: The Value of Mature-Field Replacement*. Tulsa: American Association of Petroleum Geologists Memoir 74 (2001): 63–84.

THREE **Drill, Ye Tarriers, Drill**

Bateman, A. M. *Economic Mineral Deposits*, 2nd ed. New York: John Wiley, 1950. (The Yale version of the Drake well story is on pp. 652–53.)

Deffeyes. *Beyond Oil*.

Knebel, G. M., and Guillermo Rodriguez-Eraso. "Habitat of Some Oil." *American Association of Petroleum Geologists Bulletin* 40 (1956): 547–61.

U.S. Energy Information Agency. *International Petroleum Monthly*, August 12, 2009, www.eia.doe.gov/emeu/ipsr/t11d.xls.

FOUR   The Dismal Science

Coughlin, Stephanie. Seabreeze Organic Farm, www.seabreezed.com.
Dedijer, Vladimir. *The Road to Sarajevo.* New York: Simon & Schuster, 1966.
Malkiel, Burton. G. *A Random Walk Down Wall Street.* New York: W. W. Norton, 1973.

FIVE   The Awl Bidness

Alvarez, Walter. *T. Rex and the Crater of Doom.* Princeton, NJ: Princeton University Press, 2008.
Beaton, K. *Enterprise in Oil: A History of Shell in the United States.* New York: Appleton-Century-Crofts, 1957.
Deffeyes. *Hubbert's Peak.*
Solow, Robert M. "The Economics of Resources or the Resources of Economics." *American Economic Review* 64 (1974): 1–14.
Yergin, Daniel. *The Prize: The Epic Quest for Oil, Money and Power.* New York: Free Press, 2008.

SIX   The Wider Picture

Ashton, T. S. *The Industrial Revolution, 1760–1830.* New York: Oxford University Press, 1997.
Berridge, Virginia, ed. *The Big Smoke: Fifty Years After the 1952 London Smog.* London: University of London, Institute of Historical Research, 2005.
Meadows, Donella, Dennis Meadows, Jorgen Randers, and William Behrens III. *The Limits to Growth.* New York: Universe Books, 1972.
Mohr, S. H., and G. M. Evans. "Forecasting Coal Production Until 2100." *Fuel* 88 (2009): 2059–67.
Olah, George, Alain Goeppert, and G. K. Surya Prakash. *Beyond Oil and Gas: The Methanol Economy*, 2nd ed. New York: John Wiley, 2009.

SEVEN   Uranium—the Red and the Black

Deffeyes. *Beyond Oil.* (The uranium adventure is on pp. 143–46.)
Deffeyes, Kenneth, and Ian MacGregor. "World Uranium Resources." *Scientific American* 242, no. 1 (1980): 66–76.
Deffeyes, Kenneth, Ian MacGregor, and James Kukula. *Uranium Distribution in Mined Deposits and in the Earth's Crust.* Department of Energy, Grand Junction, CO, 1978. (Copies are in the Princeton University Library.)
Hess, Harry H. *The Disposal of Radioactive Wastes on Land.* Washington, D.C.: National Academy of Science, 1957.

McPhee, John. *The Curve of Binding Energy.* New York: Farrar, Straus and Giroux, 1974.

EIGHT   Climate Change

Judson, Sheldon. "Geological and Geographical Setting (Les Eyzies)." *American School of Prehistoric Research Bulletin* 30 (1975): 19–26.
McPhee, John. "A Season on the Chalk." In *Silk Parachute.* New York: Farrar, Straus and Giroux, 2010, pp. 7–42.

NINE   Natural Gas

Deffeyes. *Hubbert's Peak.*
Frantz, Joseph H. Jr., and Valerie Jocher. "Shale Gas." Houston: Schlumberger white paper, 2005, www.unbridledenergy.com/assets/pdf/About_Shale _Gas.pdf.
McGeer, Patrick, and Enoch Durbin. *Methane: Fuel for the Future.* New York: Plenum Press, 1982.
Simmons, Matthew R. "Unlocking the Natural Gas Riddle." Houston: Simmons & Company International white paper, 2002, www.simmonscointl .com/files/GGW%20PAPER%201.pdf.

TEN   Oil from Algae

Borrell, Brendan. "Clean Dreams or Pond Scum?" *Scientific American,* July 14, 2009, www.scientificamerican.com/blog/60-second-science/post.cfm?id= clean-dreams-or-pond-scum-exxonmobi-2009-07-14.
Deffeyes, Kenneth S., and Stephen E. Deffeyes. *Nanoscale: Visualizing an Invisible World.* Cambridge, MA: MIT Press, 2009. (The structure of a lipid membrane is illustrated on pp. 50–51.)
Hamilton, Tyler. "Dow to Test Algae Ethanol." *Technology Review,* July 16, 2009, www.technologyreview.com/business/23009/?nlid=2182.

ELEVEN   Transportation Efficiency

Brown, Duncan, ed. *Effectiveness and Impact of Corporate Average Fuel Economy (CAFE) Standards.* Washington, D.C.: National Academies Press, 2002, http://books.nap.edu/openbook.php?record_id=10172&page=R2.
McPhee, John. *Uncommon Carriers.* New York: Farrar, Straus and Giroux, 2006.

TWELVE   Metal Deposits

Deffeyes, Kenneth S. "The Axial Valley: A Steady-State Feature of the Terrain." In Helgi Johnson and Bennett Smith, eds., *The Megatectonics of Continents and Oceans.* New Brunswick, NJ: Rutgers University Press, 1970, pp. 194–222.

Pauling, Linus C. *College Chemistry.* San Francisco: W. H. Freeman, 1952. (The electromotive force is discussed on p. 270.)

THIRTEEN   Recommendations

Deffeyes and Deffeyes. *Nanoscale.* (An illustrated explanation of rare-earth magnets is on pp. 114–15.)

Fagan, Brian. *The Little Ice Age: How Climate Made History.* New York: Basic Books, 2000.

French, A. P. *Vibrations and Waves.* New York: W.W. Norton, 1971. (The equations for critical damping are on pp. 69, 70.)

Index

Shell, 7, 19, 23, 45, 49, 53–54, 67
short sales, 133
Siljan Ring, 34
Silliman, Benjamin, 34, 36
silver, 119
Simmons, Matthew, 20, 97, 109, 129
Singapore, 122
Sinopec, 54
Smalley, Richard, 135
Sneider, Robert, 24
soldering alloys, 122
Solow, Robert, 52
specialist's book, 133
Spitzer, Lyman, 63
Standard Oil Company breakup, 45
Standard Oil of New Jersey, 53
steam, 46, 67
  injection, 48
Steiner, Benjamin, xv
Stendhal, 76, 78
stock market volatility, 131
straits tin, 122
sulfur dioxide, 66
sustainable resources, xvi
sustainable steady-state condition,
  136
SUVs, 40, 43, 118
Synthetic Genomics, 113

Tambora volcano, Indonesia, 94
tar sands, 46–48
Taylor, Ted, 72
telethermal ore deposits, 74
Tethys seaway, 88
Texaco, 54, 67
Texas Railroad Commission, 97
TGV, French trains, 116
third world, 66
Three Mile Island meltdown, 79
tight sands, 99

tin
  depletion versus oil depletion,
    122
  rarity of ore deposits of, 121
  veins in Cornwall, England, 121
  versus copper in Earth's crust,
    121
tin cans, 122
Titusville, Pennsylvania, 35
Total, 54
total oil, 4, 10, 25
tractor-trailer, 114
trade deficits, 42
transport efficiency, xviii, 115–18
transport systems, efficiency list,
  116
trends, 14
Tri-State lead-zinc mines, 123

Uncommon Carriers, 117
uranium, xvi
  fission, 63
  high-grade, 72
  supply, 71
uranium guzzler, 72

Van Houten, Franklyn, 85
Venezuela, xv
Ventor, Craig, 113
Ventura oil field, 37–38
Verhulst, Pierre-François, 10–11
victory gardens, World War II, 44

"wastebasket" strategy, 124–25
water, oxidizing-reducing, 76
weeds, 10